ENVIRONMENTAL SCIENCE, ENGINEERING AND TECHNOLOGY SERIES

PIPELINES FOR CARBON SEQUESTRATION: BACKGROUND AND ISSUES

ENVIRONMENTAL SCIENCE, ENGINEERING AND TECHNOLOGY SERIES

Nitrous Oxide Emissions Research Progress
Adam I. Sheldon and Edward P. Barnhart (Editors)
2009. ISBN: 978-1-60692-267-5

Fundamentals and Applications of Biosorption Isotherms, Kinetics and Thermodynamics
Yu Liu and Jianlong Wang (Editors)
2009. ISBN: 978-1-60741-169-7

Environmental Effects of Off-Highway Vehicles
Douglas S. Ouren, Christopher Haas, Cynthia P. Melcher, Susan C. Stewart, Phadrea D. Ponds, Natalie R. Sexton, Lucy Burris, Tammy Fancher and Zachary H. Bowen
2009. ISBN: 978-1-60692-936-0

Agricultural Runoff, Coastal Engineering and Flooding
Christopher A. Hudspeth and Timothy E. Reeve (Editors)
2009. ISBN: 978-1-60741-097-3

Agricultural Runoff, Coastal Engineering and Flooding
Christopher A. Hudspeth and Timothy E. Reeve (Editors)
2009. ISBN: 978-1-60876-608-6 (Online Book)

Conservation of Natural Resources
Nikolas J. Kudrow (Editor)
2009. ISBN: 978-1-60741-178-9

Conservation of Natural Resources
Nikolas J. Kudrow (Editor)
2009. ISBN: 978-1-60876-642-6 (Online Book)

Directory of Conservation Funding Sources for Developing Countries: Conservation Biology, Education and Training, Fellowships and Scholarships
Alfred O. Owino and Joseph O. Oyugi
2009. ISBN: 978-1-60741-367-7

Forest Canopies: Forest Production, Ecosystem Health and Climate Conditions
Jason D. Creighton and Paul J. Roney (Editors)
2009. ISBN: 978-1-60741-457-5

Soil Fertility
Derek P. Lucero and Joseph E. Boggs (Editors)
2009. ISBN: 978-1-60741-466-7

The Amazon Gold Rush and Environmental Mercury Contamination
Daniel Marcos Bonotto
and Ene Glória da Silveira
2009. ISBN: 978-1-60741-609-8

Buildings and the Environment
Jonas Nemecek and Patrik Schulz (Editors)
2009. ISBN: 978-1-60876-128-9

Tree Growth: Influences, Layers and Types
Wesley P. Karam (Editor)
2009. ISBN: 978-1-60741-784-2

Syngas: Production Methods, Post Treatment and Economics
Adorjan Kurucz and Izsak Bencik (Editors)
2009. ISBN: 978-1-60741-841-2

Syngas: Production Methods, Post Treatment and Economics
Adorjan Kurucz and Izsak Bencik (Editors)
2009. ISBN: 978-1-61668-214-9 (Online Book)

Process Engineering in Plant-Based Products
Hongzhang Chen
2009. ISBN: 978-1-60741-962-4

Carbon Capture and Storage including Coal-Fired Power Plants
Todd P. Carington (Editor)
2009. ISBN: 978-1-60741-196-3

Potential of Activated Sludge Utilization
Xiaoyi Yang
2010. ISBN: 978-1-60876-019-0

Psychological Approaches to Sustainability: Current Trends in Theory, Research and Applications
Victor Corral-Verdugo, Cirilo H. Garcia-Cadena and Martha Frias-Armenta (Editors)
2010. ISBN: 978-1-60876-356-6

Fluid Waste Disposal
Kay W. Canton (Editor)
2010. ISBN: 978-1-60741-915-0

Recent Progress on Earthquake Geology
Pierpaolo Guarnieri (Editor)
2010. ISBN: 978-1-60876-147-0

Check Dams, Morphological Adjustments and Erosion Control in Torrential Streams
Carmelo Consesa Garcia and Mario Aristide Lenzi (Editors)
2010. ISBN: 978-1-60876-146-3

Freshwater Ecosystems and Aquaculture Research
Felice De Carlo and Alessio Bassano (Editors)
2010. ISBN: 978-1-60741-707-1

Grassland Biodiversity: Habitat Types, Ecological Processes and Environmental Impacts
Johan Runas and Theodor Dahlgren (Editors)
2010. ISBN: 978-1-60876-542-3

Biodiversity Hotspots
Vittore Rescigno and Savario Maletta (Editors)
2010. ISBN: 978-1-60876-458-7

Handbook of Environmental Policy
Johannes Meijer and Arjan der Berg (Editors)
2010. ISBN: 978-1-60741-635-7

Environmental Modeling with GPS
Lubos Matejicek
2010. ISBN: 978-1-60876-363-4

Handbook on Agroforestry: Management Practices and Environmental Impact
Lawrence R. Kellimore (Editor)
2010. ISBN: 978-1-60876-359-7

Pipelines for Carbon Sequestration: Background and Issues
Elvira S. Hoffmann (Editor)
2010. ISBN: 978-1-60741-383-7

ENVIRONMENTAL SCIENCE, ENGINEERING AND TECHNOLOGY SERIES

PIPELINES FOR CARBON SEQUESTRATION: BACKGROUND AND ISSUES

ELVIRA S. HOFFMANN
EDITOR

Nova Science Publishers, Inc.
New York

For permission to use material from this book please contact us:
Telephone 631-231-7269; Fax 631-231-8175
Web Site: http://www.novapublishers.com

Library of Congress Cataloging-in-Publication Data

Pipelines for carbon sequestration : background and issues / editor, Elvira S. Hoffmann.
p. cm.
Includes index.
ISBN 978-1-60741-383-7 (hardcover)
1. Atmospheric carbon dioxide--Storage. 2. Geological carbon sequestration. 3. Gas pipelines. 4. Carbon dioxide mitigation. I. Hoffmann, Elvira S.
TD885.5.C3P385 2009
628.5'3--dc22
2009038415

Published by Nova Science Publishers, Inc. ✦ New York

Chapter Sources

The following chapters have been previously published:

Chapter 1 – This is an edited, excerpted and augmented edition of a United States Congressional Research Service publication, Report Order Code RL33971, dated January 17, 2008.

Chapter 2 – This is an edited, excerpted and augmented edition of a United States Congressional Research Service publication, Report Order Code RL34316, dated January 10, 2008.

Chapter 3 – This is an edited, excerpted and augmented edition of a United States Congressional Research Service publication, Report Order Code RL34307, dated April 15, 2008.

Chapter 4 – These remarks were delivered as Statements before the United States Senate, Committee on Energy and Natural Resources, dated January 31, 2007.

Contents

PREFACE

This book explores policies that Congress is considering to reduce U.S. emissions of greenhouse gases. Prominent among these policies are those promoting the capture and direct sequestration of carbon dioxide (CO2) from man-made sources such as electric power plants and manufacturing facilities. Carbon capture and sequestration is of great interest because potentially large amounts of CO2 produced by the industrial burning of fossil fuels could be sequestered. Although they are still under development, carbon capture technologies may be able to remove up to 95% of CO2 emitted from an electric power plant or other industrial source. Carbon capture and sequestration (CCS) is a three-part process involving a CO2 source facility, a long-term CO2 sequestration site, and an intermediate mode of CO2 transportation - typically pipelines.This book consists of public documents which have been located, gathered, combined, reformatted, and enhanced with a subject index, selectively edited and bound to provide easy access.

Chapter 1 - Congress is examining potential approaches to reducing manmade contributions to global warming from U.S. sources. One approach is carbon capture and sequestration (CCS) — capturing CO_2 at its source (e.g., a power plant) and storing it indefinitely (e.g., underground) to avoid its release to the atmosphere. A common requirement among the various techniques for CCS is a dedicated pipeline network for transporting CO_2 from capture sites to storage sites.

In the 110th Congress, there has been considerable debate on the capture and sequestration aspects of carbon sequestration, while there has been relatively less focus on transportation. Nonetheless, there is increasing understanding in Congress that a national CCS program could require the construction of a substantial network of interstate CO_2 pipelines. S. 2144 and S. 2191 would require the Secretary of Energy to study the feasibility of constructing and operating such a network of pipelines. S. 2323 would require carbon sequestration projects to evaluate the most cost-efficient ways to integrate CO_2 sequestration, capture, and transportation. S. 2149 would allow seven-year accelerated depreciation for qualifying CO_2 pipelines. P.L. 110-140, signed by President Bush on December 19, 2007, requires the Secretary of the Interior to recommend legislation to clarify the issuance of CO_2 pipeline rights-of-way on public land.

That CCS and related legislation have been more focused on the capture and storage of CO_2 than on its transportation, reflects a perception that transporting CO_2 via pipelines does not present a significant barrier to implementing large-scale CCS. Notwithstanding this perception, and even though regional CO_2 pipeline networks already operate in the United States for enhanced oil recovery (EOR), developing a more expansive national CO_2 pipeline network for CCS could pose numerous new regulatory and economic challenges. There are important unanswered questions about pipeline network requirements, economic regulation, utility cost recovery, regulatory classification of CO_2 itself, and pipeline safety. Furthermore, because CO_2 pipelines for EOR are already in use today, policy decisions affecting CO_2 pipelines take on an urgency that is, perhaps, unrecognized by many. Federal classification of CO_2 as both a commodity (by the Bureau of Land Management) and as a pollutant (by the Environmental Protection Agency) could potentially create an immediate conflict which may need to be addressed not only for the sake of future CCS implementation, but also to ensure consistency of future CCS with CO_2 pipeline operations today.

In addition to these issues, Congress may examine how CO_2 pipelines fit into the nation's overall strategies for energy supply and environmental protection. If policy makers encourage continued consumption of fossil fuels under CCS, then the need to foster the other energy options may be diminished — and vice versa. Thus decisions about CO_2 pipeline infrastructure could have consequences for a broader array of energy and environmental policies.

Chapter 2 - Congress is considering policies promoting the capture and sequestration of carbon dioxide (CO_2) from sources such as electric power plants. Carbon capture and sequestration (CCS) is a process involving a CO_2 source facility, a long-term CO_2 sequestration site, and CO_2 pipelines. There is an increasing perception in Congress that a national CCS program could require the construction of a substantial network of interstate CO_2 pipelines. However, divergent views on CO_2 pipeline requirements introduce significant uncertainty into overall CCS cost estimates and may complicate the federal role, if any, in CO_2 pipeline development. S. 2144 and S. 2191 would require the Secretary of

Energy to study the feasibility of constructing and operating such a network of pipelines. S. 2323 would require carbon sequestration projects to evaluate the most cost-efficient ways to integrate CO_2 sequestration, capture, and transportation. P.L. 110-140, signed by President Bush on December 19, 2007, requires the Secretary of the Interior to recommend legislation to clarify the issuance of CO_2 pipeline rights-of-way on public land.

The cost of CO_2 transportation is a function of pipeline length and other factors. This chapter examines key uncertainties in CO_2 pipeline requirements for CCS by contrasting hypothetical pipeline scenarios for 11 major coal-fired power plants in the Midwest Regional Carbon Sequestration Partnership region. The scenarios illustrate how different assumptions about sequestration site viability can lead to a 20-fold difference in CO_2 pipeline lengths, and, therefore, similarly large differences in capital costs. From the perspective of individual power plants, or other CO_2 sources, variable costs for CO_2 pipelines may have significant ramifications. If CO_2 pipeline costs for specific regions reach tens, or even hundreds, of millions of dollars per plant, then power companies may have difficulty securing the capital financing or regulatory approval needed to construct or retrofit fossil fuel-powered plants in these regions. High CO_2 transportation costs also could increase electricity prices in "sequestration-poor" regions relative to regions able to sequester CO_2 more locally.

As CO_2 pipelines get longer, the state-by-state siting approval process may become complex and protracted, and may face public opposition. Because CO_2 pipeline requirements in a CCS scheme are driven by the relative locations of CO_2 sources and sequestration sites, identification and validation of such sites must explicitly account for CO_2 pipeline costs if the economics of those sites are to be fully understood. Since transporting CO_2 to distant locations can impose significant additional costs to a facility's carbon control infrastructure, facility owners may seek regulatory approval for as many sequestration sites as possible and near to as many facilities as possible. If CCS moves to widespread implementation, government agencies and private companies may face challenges in identifying, permitting, developing, and monitoring the large number of localized sequestration reservoirs that may be proposed. However, even as viable sequestration reservoirs are being identified, it is unclear which CO_2 source facilities will have access to them, under what time frame, and under what conditions. Given the potential size of a national CO_2 pipelines network, many billions of dollars of capital investment may be affected by policy decisions made today.

Chapter 3 - In the last few years there has been a surge in interest in carbon capture and sequestration (CCS) as a way to mitigate manmade carbon dioxide

(CO_2) emissions and thereby help address climate change concerns. This approach may require the transportation of captured carbon dioxide via pipeline to an underground storage reservoir with the intent of keeping it out of the atmosphere. Congress has begun to address this issue.

The recently enacted Energy Independence and Security Act of 2007 (P.L. 110-140) contains measures to promote research and development of CCS technology, to assess sequestration capacity, and to clarify the framework for issuance of CO_2 pipeline rights-of-way on public land. Other legislative measures, including S. 2191 and S. 2323, have also sought to encourage the development of CO_2 sequestration, capture, and transportation technology. Further, measures have been introduced in Congress, including the Carbon Dioxide Pipeline Study Act of 2007 (S. 2144), that would require the Secretary of Energy to study the feasibility of constructing and operating a network of CO_2 pipelines for CCS. If these pipelines crossed state lines, it could raise important issues concerning regulatory jurisdiction over siting and pricing.

This chapter discusses federal jurisdictional uncertainty over CO_2 pipelines under existing law, as well as potential legislative activity to address any possible regulatory "gap" in jurisdiction. It should be noted that such a potential gap does not necessarily demand federal legislative resolution. Nevertheless, some analysts, drawing on the history of oil and natural gas pipeline development, anticipate a potential need for better defined federal legislative/regulatory authority over an interstate CO_2 pipeline network. Accordingly, Congress may be called upon to consider whether existing federal jurisdictional disclaimers and the state-by-state regulatory structure for Enhanced Oil Recovery (EOR) pipelines is an appropriate regulatory scheme for a possible interstate CO_2 pipeline network in support of CCS. Congress could opt to amend existing statutes, possibly those related to the interstate pipeline jurisdiction of the Federal Energy Regulatory Commission (FERC) or the Surface Transportation Board (STB) to provide for definitive CO_2 pipeline rate jurisdiction by the federal government if it does not favor the existing regulatory scheme. Alternatively, Congress could establish another federal regulator of CO_2 pipelines or possibly other legislation to amend the existing regulatory scheme.

Chapter 4 -Regulatory Aspects Of Carbon Capture, Transportation & Sequestration Hearing before the United States Senate Committee on Energy and Natural Resources.

In: Pipelines for Carbon Sequestration
Editors: Elvira S. Hoffmann

ISBN: 978-1-60741-383-7

Chapter 1

CARBON DIOXIDE (CO_2) PIPELINES FOR CARBON SEQUESTRATION: EMERGING POLICY ISSUES

***Paul W. Parformak*[1] *and Peter Folger*[2]**

[1] Specialist in Energy and Infrastructure Policy Resources, Science, and Industry Division

[2] Specialist in Energy and Natural Resources Policy Resources, Science, and Industry Division

SUMMARY

Congress is examining potential approaches to reducing manmade contributions to global warming from U.S. sources. One approach is carbon capture and sequestration (CCS) — capturing CO_2 at its source (e.g., a power plant) and storing it indefinitely (e.g., underground) to avoid its release to the atmosphere. A common requirement among the various techniques for CCS is a dedicated pipeline network for transporting CO_2 from capture sites to storage sites.

In the 110th Congress, there has been considerable debate on the capture and sequestration aspects of carbon sequestration, while there has been relatively less focus on transportation. Nonetheless, there is increasing understanding in Congress that a national CCS program could require the construction of a substantial network of interstate CO_2 pipelines. S. 2144 and S. 2191 would require the Secretary of Energy to study the feasibility of constructing and operating such a network of pipelines. S. 2323 would require carbon sequestration projects to evaluate the most cost-efficient ways to integrate CO_2 sequestration, capture, and transportation. S. 2149 would allow seven-year accelerated depreciation for qualifying CO_2 pipelines. P.L. 110-140, signed by President Bush on December 19, 2007, requires the Secretary of the Interior to recommend legislation to clarify the issuance of CO_2 pipeline rights-of-way on public land.

That CCS and related legislation have been more focused on the capture and storage of CO_2 than on its transportation, reflects a perception that transporting CO_2 via pipelines does not present a significant barrier to implementing large-scale CCS. Notwithstanding this perception, and even though regional CO_2 pipeline networks already operate in the United States for enhanced oil recovery (EOR), developing a more expansive national CO_2 pipeline network for CCS could pose numerous new regulatory and economic challenges. There are important unanswered questions about pipeline network requirements, economic regulation, utility cost recovery, regulatory classification of CO_2 itself, and pipeline safety. Furthermore, because CO_2 pipelines for EOR are already in use today, policy decisions affecting CO_2 pipelines take on an urgency that is, perhaps, unrecognized by many. Federal classification of CO_2 as both a commodity (by the Bureau of Land Management) and as a pollutant (by the Environmental Protection Agency) could potentially create an immediate conflict which may need to be addressed not only for the sake of future CCS implementation, but also to ensure consistency of future CCS with CO_2 pipeline operations today.

In addition to these issues, Congress may examine how CO_2 pipelines fit into the nation's overall strategies for energy supply and environmental protection. If policy makers encourage continued consumption of fossil fuels under CCS, then the need to foster the other energy options may be diminished — and vice versa. Thus decisions about CO_2 pipeline infrastructure could have consequences for a broader array of energy and environmental policies.

INTRODUCTION

Congress has long been concerned about the impact of global climate change that may be caused by manmade emissions of carbon dioxide (CO_2) and other greenhouse gases.[1] Congress is also debating policies related to global warming and is examining a range of potential initiatives to reduce manmade contributions to global warming from U.S. sources.[2] One approach to mitigating manmade greenhouse gas emissions is direct sequestration: capturing CO_2 at its source, transporting it via pipelines, and storing it indefinitely to avoid its release to the atmosphere.[3] This paper explores one component of direct sequestration — transporting CO_2 in pipelines.

Carbon capture and storage (CCS) is of great interest because potentially large amounts of CO_2 emitted from the industrial burning of fossil fuels in the United States could be suitable for sequestration. Carbon capture technologies can potentially remove 80%-95% of CO_2 emitted from an electric power plant or other industrial source. Power plants are the most likely initial candidates for CCS because they are predominantly large, single-point sources, and they contribute approximately one-third of U.S. CO_2 emissions from fossil fuels.

There are many technological approaches to CCS. However, one common requirement for nearly all large-scale CCS schemes is a system for transporting CO_2 from capture sites (e.g., power plants) to storage sites (e.g., underground reservoirs). Transporting captured CO_2 in relatively limited quantities is possible by truck, rail, and ship, but moving the enormous quantities of CO_2 implied by a widespread implementation of CCS technologies would likely require a dedicated interstate pipeline network.

In the 110th Congress, there has been considerable debate on the capture and sequestration aspects of carbon sequestration, while there has been relatively less focus on transportation. Nonetheless, there is increasing understanding in Congress that a national CCS program could require the construction of a substantial network of interstate CO_2 pipelines. The Carbon Dioxide Pipeline Study Act of 2007 (S. 2144), introduced by Senator Coleman and nine cosponsors on October 4, 2007, would require the Secretary of Energy to study the feasibility of constructing and operating such a network of CO_2 pipelines. The America's Climate Security Act of 2007 (S. 2191), introduced by Senator Lieberman and nine cosponsors on October 18, 2007, and reported out of the Senate Environment and Public Works Committee in amended form on December 5, 2007, contains similar provisions (Sec. 8003). The Carbon Capture and Storage Technology Act of 2007 (S. 2323), introduced by Senator Kerry and one cosponsor on November 7, 2007, would require carbon sequestration projects authorized by the act to evaluate the most cost-efficient ways to integrate CO_2 sequestration, capture, and transportation (Sec. 3(b)(5)). The Coal Fuels and Industrial Gasification Demonstration and Development Act of 2007 (S.2149) introduced by Senator Dorgan on October 4, 2007, would allow accelerated depreciation for certain new CO_2 pipelines. The Energy Independence and Security Act of 2007 (P.L. 110-140) signed by President Bush, as amended, on December 19, 2007, requires the Secretary of the Interior to recommend legislation to clarify the appropriate framework for issuing CO_2 pipeline rights-of-way on public land (Sec. 714(7)).

Legislative focus on the capture and storage components of direct carbon sequestration reflects a perception that transporting CO_2 via pipelines does not present a significant barrier to implementing large-scale CCS. Even though regional CO_2 pipeline networks alreadyoperate in the United Statesfor enhanced oil recovery (EOR), developing a more expansive national CO_2 pipeline network for CCS could pose numerous new regulatory and economic challenges. As one analyst has remarked,

> Each of the individual technologies involved in the transport portion of the CCS process is mature, but integrating and deploying them on a massive scale will be a complex task. "The question is, how would the necessary pipeline network be established and evolve?"[4]

A thorough consideration of potential CCS approaches necessarily involves an assessment of their overall requirements for CO_2 transportation by pipeline, including the possible federal role in establishing an interstate CO_2 pipeline network.

This chapter introduces key policy issues related to CO_2 pipelines which may require congressional attention. It summarizes the technological requirements for CO_2 pipeline transportation under a comprehensive CCS strategy. It characterizes these requirements relative to the existing CO_2 pipeline infrastructure in the United States used for EOR. The report summarizes policy issues related to CO_2 pipeline development, including uncertainty about pipeline network requirements, economic regulation, utility cost recovery, regulatory classification of CO_2 itself, and pipeline safety. The report concludes with perspectives on CO_2 pipelines in the context of the nation's overall energy and infrastructure requirements.

BACKGROUND

Carbon sequestration policies are inextricably tied to the function and availability of the necessary technologies. Consequently, discussion of CCS policy alternatives benefits from a basic understanding of the physical processes involved, and relevant experience with existing infrastructure. This section provides a basic overview of carbon sequestration processes overall, as well as specific U.S. experience with CO_2 pipelines.[5]

Carbon Capture and Sequestration

Carbon capture and sequestration is essentially a three-part process involving a CO_2 source facility, a long-term CO_2 storage site, and an intermediate mode of CO_2 transportation.

Capture

The first step in direct sequestration is to produce a concentrated stream of CO_2 for transport and storage. Currently, three main approaches are available to capture CO_2 from large-scale industrial facilities or power plants:

- **pre-combustion,** which separates CO_2 from fuels by combining them with air and/or steam to produce hydrogen for combustion and CO_2 for storage,
- **post-combustion**, which extracts CO_2 from flue gases following combustion of fossil fuels or biomass, and
- **oxyfuel combustion**, which uses oxygen instead of air for combustion, producing flue gases that consist mostly of CO_2 and water from which the CO_2 is separated.[6]

These approaches vary in terms of process technology and maturity, but all yield a stream of extracted CO_2 which may then be compressed to increase its density and make it easier (and cheaper) to transport. Although technologies to separate and compress CO_2 are commercially available, they have not been applied to large-scale CO_2 capture from power plants for the purpose of long-term storage.[7]

Transportation

Pipelines are the most common method for transporting large quantities of CO_2 over long distances. CO_2 pipelines are operated at ambient temperature and high pressure, with primary compressor stations located where the CO_2 is injected and booster compressors located as needed further along the pipeline.[8] In overall construction, CO_2 pipelines are similar to natural gas pipelines, requiring the same attention to design, monitoring for leaks, and protection against overpressure, especially in populated areas.[9] Many analysts consider CO_2 pipeline technology to be mature, stemming from its use since the 1970s for EOR and in other industries.[10] Marine transportation may also be feasible when CO_2 needs to be transported over long distances or overseas; however, many manmade CO_2 sources arelocatedfar from navigablewaterways, so such a scheme would still likely require pipeline construction between CO_2 sources and port terminals. Rail cars and trucks can also transport CO_2, but these modes would be logistically impractical for large-scale CCS operations.

Sequestration in Geological Formations

In most CCS approaches, CO_2 would be transported by pipeline to a porous rock formation that holds (or previously held) fluids where the CO_2 would be injected underground. When CO_2 is injected over 800 meters deep in a typical storage formation, atmospheric pressure induces the CO_2 to become relatively dense and less likely to migrate out of the formation. Injecting CO_2 into such formations uses existing technologies developed primarily for oil and natural gas production which potentially could be adapted for long-term storage and monitoring of CO_2. Other underground injection applications in practice today, such as natural gas storage, deep injection of liquid wastes, and subsurface disposal of oil-field brines, also provide potential technologies and experience for sequestering CO_2.[11] Three main types of geological formations are being considered for carbon sequestration: (1) oil and gas reservoirs, (2) deep saline reservoirs, and (3) unmineable coal seams. The overall capacity for CO_2 storage in such formations is potentially huge if all the sedimentary basins in the world are considered.[12] The suitability of any particular site, however, depends on many factors, including proximity to CO_2 sources and other reservoir-specific qualities like porosity, permeability, and potential for leakage.

Existing U.S. CO_2 Pipelines

The oldest long-distance CO_2 pipeline in the United States is the 225 kilometer Canyon Reef Carriers Pipeline (in Texas), which began service in 1972 for EOR in regional oil fields.[13] Other large CO_2 pipelines constructed since then, mostly in the Western United States, have expanded the CO_2 pipeline network for EOR. These pipelines carry CO_2 from naturally occurring underground reservoirs, atural gas processing facilities, ammonia manufacturing plants, and a large coal gasification project to oil fields. Additional pipelines may carry CO_2 from other manmade sources to supply a range of industrial applications. Altogether, approximately 5,800 kilometers (3,600 miles) of CO_2 pipeline operate today in the United States.[14]

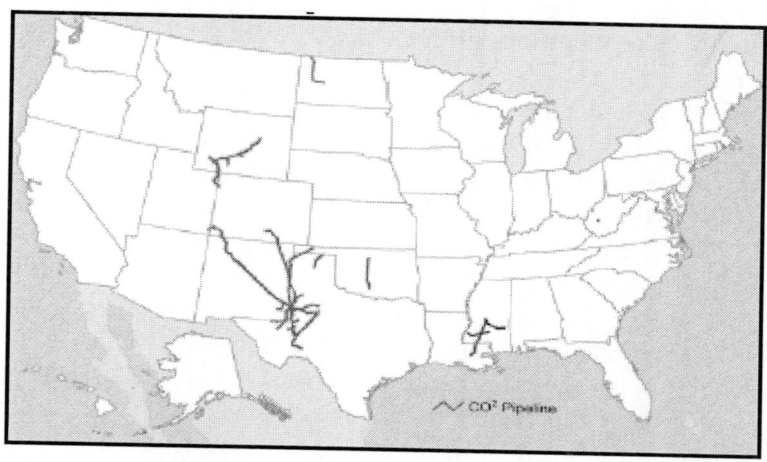

Sources: Denbury Resources Inc., "EOR: The Economic Alternative for CCS," Slide presentation (October 2007). [http://www.gasification.org/ Docs/2007_Papers/ 25EVAN.pdf]; U.S. Dept. of Transportation, National Pipeline Mapping System, Official use only. (June 2005). [https://www.npms.phmsa.dot.gov]

Figure 1. Major CO_2 Pipelines in the United States

The locations of the major U.S. CO_2 pipelines are shown in Figure 1. By comparison, nearly 800,000 kilometers (500,000 miles) of natural gas and hazardous liquid transmission pipelines crisscross the United States.[15]

KEY ISSUES FOR CONGRESS

Congressional consideration of potential CCS policies is still evolving, but so far initiatives have focused more on developing capture and sequestration technologies than on transportation. Specific legislative proposals in the 110th Congress reflect the current perception that CO_2 capture probably represents the largest technological hurdle to implementing widespread CCS, and that CO_2 transportation by pipelines does not present as significant a barrier. While these perceptions may be accurate, industry and regulatory analysts have begun to identify important policy issues related specifically to CO_2 pipelines which may require congressional attention.

CO_2 Pipeline Requirements for CCS

Although any widespread CCS scheme in the United States would likely require dedicated CO_2 pipelines, there is considerable uncertainty about the size and configuration of the pipeline network required. This uncertainty stems, in part, from uncertainty about the suitability of geological formations to sequester captured CO_2 and the proximity of suitable formations to specific sources. One recent analysis concludes that 77% of the total annual CO_2 captured from the major North American sources may be stored in reservoirs directly underlying these sources, and that an additional 18% may be stored within 100 miles of additional sources.[16] If this were the case, the need for new CO_2 pipelines would be limited to onsite transportation and a relatively small number of long-distance pipelines (only a subset of which might need to be interstate pipelines).

Other analysts suggest that captured CO_2 may need to be sequestered, at least initially, in more centralized reservoirs to reduce potential risks associated with CO_2 leaks.[17] They suggest that, given current uncertainty about the suitability of various on-site geological formations for long-term CO_2 storage, certain specific types of formations (e.g., salt caverns) may be preferred as CO_2 repositories because they have adequate capacity and are most likely to retain sequestered CO_2 indefinitely. As geologic formations are characterized in more detail and suitable repositories identified, CO_2 sources can be mapped against storage sites withincreasingcertainty. The current uncertainty over proximity of sources to storage sites, however, implies a wide range of possible pipeline configurations and a wide range of possible costs.

Whether CCS policies ultimately lead to centralized or decentralized storage configurations remains to be seen; however, pipeline requirements and storage configurations are closely related. A 2007 study at the Massachusetts Institute of Technology (MIT) concluded that "the majority of coal-fired power plants are situated in regions where there are high expectations of having CO_2 sequestration sites nearby."[18] In these cases, the MIT study estimated the cost of CO_2 transport and injection to be less than 20% of total CCS costs. However, the study also stated that the costs of CO_2 pipelines are highly non-linear with respect to the quantity transported, and highly variable due to "physical ... and political considerations."[19] Another 2007 study, at Duke University, concluded that "geologic sequestration is not economically or technically feasible within North Carolina," but "may be viable if the captured CO_2 is piped out of North Carolina and stored elsewhere."[20] There are also significant scale economies for large, integrated CO_2 pipeline networks that link many sources together rather than single, dedicated pipelines between individual sources and storage reservoirs.[21] As

Congress considers CCS policies, it may examine the relationship between CO_2 reservoir sites and pipeline requirements.

Economic Regulation

Economic regulation of interstate pipelines by the federal government is generally intended to ensure pipelinesfulfill *common carrier* obligations by charging reasonable rates; providing rates and services to all upon reasonable request; not unfairly discriminating among shippers; establishing reasonable classifications, rules, and practices; and interchanging traffic with other pipelines or transportation modes.[22] If interstate CO_2 pipelines for carbon sequestration are ultimately to be developed, it will raise important regulatory questions in this context because federal jurisdiction over hypothetical interstate CO_2 pipeline siting and rate decisions is not clear. Based on their current regulatory roles, two of the more likely candidates for jurisdiction over interstate pipelines transporting CO_2 for purposes of CCS are the Federal Energy Regulatory Commission (FERC) and the Surface Transportation Board (STB).[23] However, both agencies have at some point expressed a position that interstate CO_2 pipelines are not within their purview, as summarized below.[24]

Federal Jurisdiction over CO_2 Pipelines

The Natural Gas Act of 1938 (NGA) vests in FERC the authority to issue "certificates of public convenience and necessity" for the construction and operation of interstate natural gas pipeline facilities.[25] FERC is also charged with extensive regulatory authority over the siting of natural gas import and export facilities, as well as rates for transportation of natural gas and other elements of transportation service. FERC also has jurisdiction over regulation of oil pipelines pursuant to the Interstate Commerce Act (ICA).[26] Although FERC is not involved in the oil pipeline siting process, as with natural gas, FERC does regulate transportation rates and capacity allocation for oil pipelines.[27] Jurisdiction over rate regulation for pipelines "other" than "water, gas or oil" pipelines resides with the STB, a decisionally independent regulatory agency affiliated with the Department of Transportation.[28] The STB acts as a forum for resolution of disputes related to pipelines within its jurisdiction. Parties who wish to challenge a rate or another aspect of a pipeline's common carrier service must petition the STB for a hearing, however; there is no ongoing regulatory oversight.

Although CO_2 pipelines are not explicitly excluded from FERC jurisdiction by statute, FERC ruled in 1979 that they are not subject to the Commission's jurisdiction because they do not transport *natural gas* for heating purposes.[29] Likewise, the ICC in 1980 concluded that Congress intended to exclude all types of gas, including CO_2, from ICC regulation. After making the initial decision that it likely did not have jurisdiction over CO_2 pipelines, the ICC did conclude that the issue was "important enough to institute a proceeding and accept comments on the petition and our view on it."[30] After the comment period the ICC confirmed its view that CO_2 pipelines were excluded from the ICC's (and, therefore, the STB's) jurisdiction.[31] Thus, the two federal regulatory agencies that, generally speaking, have jurisdiction over interstate pipeline rate and capacity allocation matters appear to have rejected explicitly jurisdiction over CO_2 siting and rates, and there is no legislative or judicial history to suggest that their rejections were improper at the time. Absent federal authority, CO_2 pipelines are regulated to varying degrees by the states.

Potential Issues Related to ICC Jurisdiction

Notwithstanding the ICC's 1980 disclaimer of jurisdiction over CO_2 pipelines, other evidence indirectly suggests the possibility that interstate CO_2 pipelines could still be considered subject to STB jurisdiction. For example, an April 1998 report by the General Accounting Office (GAO)[32] stated that interstate CO_2 pipelines, as well as pipelines transporting other gases are subject to the board's oversight authority. The STB reviewed the GAO's analysis and, apparently, did not object to this jurisdictional classification.[33] Furthermore, although the STB is the successor to the now-defunct ICC, the STB conceivably could determine that its jurisdiction is not governed by the ICC's decision in the CO_2 matter. Indeed, the Supreme Court has ruled that federal agencies are not precluded from changing their positions on the issue of regulatory jurisdiction. According to the Court, "an initial agency interpretation is not instantly carved in stone. On the contrary, the agency, to engage in informed rulemaking, must consider varying interpretations and the wisdom of its policy on a continuing basis."[34] Accordingly, regulation of CO_2 pipelines for CCS purposes by the STB (or by FERC, for that matter) under existing statutes remains a possibility.

Policy Implications for Rate Regulation

If CCS technology develops to the point where interstate CO_2 pipelines become more common, and if FERC and the STB continue to disclaim jurisdiction over CO_2 pipelines, then the absence of federal regulation described above may pose policy challenges. In particular, with many more pipeline users and interconnections than exist today, complex common carrier issues might arise.[35] One potential concern, for example, is whether rates should be set separately for existing pipelines carrying CO_2 as a valuable commercial commodity (e.g., for EOR), versus new pipelines carrying CO_2 as industrial pollution for disposal. Furthermore, if rates are not reviewed prior to pipeline construction, it might be difficult for regulators to ensure the reasonableness of CO_2 pipeline rates until after the pipelines were already in service. If CO_2 pipeline connections become mandatory under future regulations, such arrangements might expose pipeline users to abuses of potential market power in CO_2 pipeline services, at least until rate cases could be heard. Presiding over a large number of CO_2 rate cases of varying complexity in a relatively short time frame might also be administratively overwhelming for state agencies, which may have limited resources available for pipeline regulatory activities.

Siting Authority

A company seeking to construct a CO_2 pipeline must secure siting approval from the relevant regulatory authorities and must subsequently secure rights of way from landowners along the pipeline right by purchasing easements or by eminent domain. However, since federal agencies claim no regulatory authority with respect to CO_2 pipeline construction, potential builders of new CO_2 pipelines do not require, and could not obtain, federal approval to construct new pipelines. Likewise, federal regulators claim no eminent domain authority for pipeline construction, and so cannot ensure that pipeline companies can secure rights of way to construct new pipelines. By contrast, companies seeking to build interstate *natural gas* pipelines must first obtain certificates of public convenience and necessity from FERC under the Natural Gas Act (15 U.S.C. §§ 717, et seq.). Such certification may include safety and security provisions with respect to pipeline routing, safety standards and other factors.[36] A certificate of public convenience and necessity granted by FERC (15 U.S.C. § 717f(h)) confers eminent domain authority.

The state-by-state siting approval process for CO_2 pipelines may be complex and protracted, and may face public opposition, especially in populated or environmentally sensitive areas. As the National Commission on Energy Policy (NCEP) states in its 2006 report:[37]

> Recent developments notwithstanding, most new energy projects are still regulated primarily at the state level and public opposition remains inextricably intertwined with local concerns, including environmental and ecosystem impacts as well as, in some cases, complex issues of property rights and competing land uses.... In some cases, upstream or downstream infrastructure requirements — such as the need for ... underground carbon sequestration sites ... may generate as

> much if not more opposition than the energy facilities they support. At the same time — and despite recent moves toward consolidated oversight by FERC or other regulatory authorities — fragmented permitting processes, nonstandard permitting requirements, and interagency redundancy often still compound siting challenges.

Securing rights of way along existing easements for other infrastructure (e.g., natural gas pipelines, electric transmission lines) may be one way to facilitate the siting of new CO_2 pipelines. However, existing easements may be ambiguous as to the right of the easement holder to install and operate CO_2 pipelines. Questions may also arise as to compensation for landowners or easement holders for use of such easements, and as to whether existing easements can be sold or leased to CO_2 pipeline companies.[38] A related issue is whether state condemnation laws, which are often used to secure sites for infrastructure deemed to be in the public interest, allow for CO_2 pipelines to be treated as public utilities or common carriers. This issue also arises on federal lands managed by the Bureau of Land Management (BLM). New CO_2 pipelines through BLM lands potentially may be sited under right of way provisions in either the Federal Land Policy and Management Act (FLPMA; 43 U.S.C. § 35) or the Mineral Leasing Act (MLA; 30 U.S.C. § 185). However, the MLA imposes a common carrier requirement while the FLPMA does not. Although the agency currently permits CO_2 pipelines for EOR under the MLA,[39] CO_2 pipeline companies seeking to avoid common carrier requirements under CCS schemes may litigate to secure rights of way under FLPMA.[40] Provisions in P.L. 110-140 require the Secretary of the Interior to recommend legislation to clarify the appropriate framework for issuing CO_2 pipeline rights-of-way on federal land (Sec. 714(7)).

Another complicating factor in the siting of CO_2 pipelines for CCS is the types of locations of existing CO_2 sources. Although a network of long-distance CO_2 pipelines exists in the United States today for EOR, these pipelines are sited mostly in remote areas accustomed to the presence of large energy infrastructure. However, many potential sources of CO_2, such as power plants, are located in populated regions, manywith a history of public resistance to the siting of energy infrastructure. If a widespread CO_2 pipeline network is required to support CCS, the ability to site pipelines to serve such facilities may become an issue requiring congressional attention. As the NCEP concluded, “In sum, it seems probable that the siting of critical infrastructure will continue to present a major challenge for policymakers.”[41]

Commodity vs. Pollutant Classification

Under a comprehensive CCS policy, captured CO_2 arguably could be classified as either a commodity or as a pollutant. CO_2 used in EOR is considered to be a commodity, and is regulated as such by the states. Because captured CO_2 may be sold as a valuable commodity for EOR, and may have further economic potential for enhanced recovery of coal bed methane (ECBM), some argue that all CO_2 under a CCS scheme should be classified as a commodity.[42] However, it is unlikely that the quantities of CO_2 captured under a widely implemented CCS policy could all be absorbed in EOR or ECBM applications. In the long run, significant quantities of captured CO_2 will have to be disposed as industrial pollution, with negative economic value.[43] Furthermore, on April 2, 2007, the U.S. Supreme Court held that the Clean Air Act gives the U.S. Environmental Protection Agency (EPA) the authority to regulate greenhouse gas emissions, including CO_2, from new motor vehicles.[44] The court also held that EPA cannot interpose policy considerations to refuse to exercise this authority. While the specifics of EPA regulation under this ruling might be subject to agency discretion, it has implications for the regulation of CO_2 emissions from stationary sources, such as power plants.

Separately, EPA has also concluded that geologic sequestration of captured CO_2 through well injection meets the definition of "underground injection" in § 1421(d)(1) of the Safe Drinking Water Act (SDWA).[45] EPA anticipates protecting underground sources of drinking water, through its authority under the SDWA, from "potential endangerment" as a result of underground injection of CO_2 in anticipated CCS pilot projects. EPA's assertion of authority under SDWA for underground injection of CO_2 during CCS pilot studies may contribute to uncertainty over future classification of CO_2 as a commodity or a pollutant.

Conflicting classification of captured CO_2 as either a commodity or pollutant has important implications for CO_2 pipeline development. For example, classifying all CO_2 as a pollutant not only would contradict current state and BLM treatment of CO_2 for EOR, but might also undermine an interstate commerce rationale for FERC regulation of CO_2 pipelines. On the other hand, classifying all CO_2 as a commodity would create other policy contradictions, for example, in regions like New England where EOR may be impracticable. Under either scenario, legislative and regulatory ambiguities would arise — especially for an integrated, interstate CO_2 pipeline network carrying a mixture of "commodity" CO_2 and "pollutant" CO_2. Resolving these ambiguities to establish a consistent and workable CCS policy could likely be an issue for Congress.

Pipeline Costs

If an extensive network of pipelines is required for CO_2 transportation, pipeline costs may be a major consideration in CCS policy. MIT estimated overall annualized pipeline transportation (and storage) costs of approximately $5 per metric ton of CO_2.[46] If CO_2 sequestration rates in the United States were on the order of 1 billion metric tons per year at mid-century, as some analysts propose, annualized pipeline costs would run into the billions of dollars. Furthermore, because most pipeline costs are initial capital costs, up-front capital outlays for a new CO_2 pipeline network would be enormous. The 2007 Duke study, for example, estimated it would cost approximately $5 billion to construct a CO_2 trunk line along existing pipeline rights of wayto transport captured CO_2 from North Carolina to potential sequestration sites in the Gulf states and Appalachia.[47] Within the context of overall CO_2 pipeline costs, several specific cost-related issues may warrant further examination by Congress.

Materials Costs

Analysts commonly develop cost estimates for CO_2 pipelines based on comparable construction costs for natural gas pipelines, and to a lesser extent, petroleum product pipelines. In most cases, these comparisons appear appropriate since CO_2 pipelines are similar in design and operation to other pipelines, especially natural gas pipelines. A University of California (UC) study analyzing the costs of U.S. transmission pipelines constructed between 1991 and 2003 found that, on average, labor accounted for approximately 45% of the total construction costs. Materials, rights of way, and miscellaneous costs accounted for 26%, 22%, and 7% of total costs, respectively.[48] Materials cost was most closely dependent upon pipeline size, accounting for an increasing fraction of the total cost with increasing pipeline size, from 15% to 35% of total costs. The MIT study estimated that transportation of captured CO_2 from a 1 gigawatt coal-fired power plant would require a pipe diameter of 16 inches.[49] According to the UC analysis, total construction costs for such a pipe between 1991 and 2003 averaged around $800,000 per mile (in 2002 dollars), although the study stated that costs for any individual pipeline could vary by a factor of five depending its location.[50]

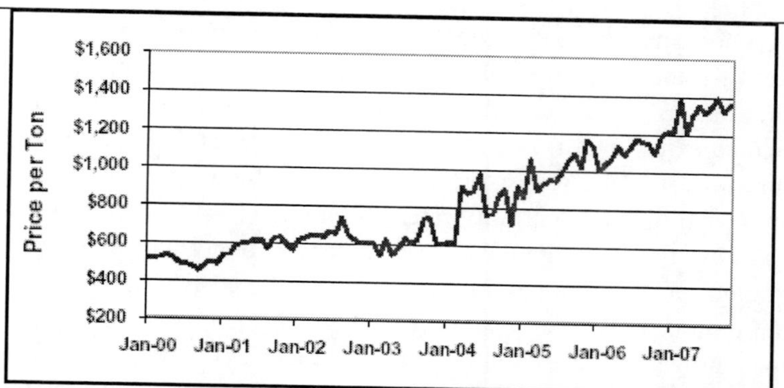

Source: *Preston Pipe & Tube Report*. Pipe prices represent average transaction price (by weighted average value) for double-submerged arc-welded pipe > 24" diameter, combining domestic and import shipments. Prices are reported through October 2007.

Figure 2. U.S. Prices for Large Diameter Steel Pipe

Since pipeline materials make up a significant portion of CO_2 pipeline construction costs, analysts have called attention to rising pipeline materials costs, especially steel costs, as a concern for policymakers.[51] Following a period of low steel prices and company bankruptcies earlier in the decade, the North American steel industry has returned to profitability and enjoys strong domestic and global demand.[52] Now, higher prices resulting from both strong demand and increased production costs for carbon steel plate, used in making large-diameter pipe, may alter the basic economics of CO_2 pipeline projects and CCS schemes overall. As Figure 2 shows, the price of large-diameter pipe was generally around $600 per ton in late 2001 and early 2002. By late 2007, the price of pipe was approaching $1,400 per ton. Analysts forecast carbon steel prices to decline over the next two years, but only gradually, and to a level still more than double the price early in the decade.[53]

If some form of CCS is effectively mandated in the future, a surge in demand for new CO_2 pipe, in competition with demand for natural gas and oil pipelines, may exacerbate the trend of rising prices for pipeline materials, and could even lead to shortages of pipe steel from North American sources. As a consequence, the availability and cost of pipeline steel to build such a CO_2 pipeline network for CCS may be a limiting factor for widespread CCS implementation.

Cost Recovery

In states where traditional rate regulation exists, construction and operation of CO_2pipelines for CCS could raise questions about cost recovery for electric utilities under state utility regulation. If, for example, a CO_2 pipeline is constructed for the exclusive use of a single power plant for on-site (or nearby) CO_2 sequestration, and is owned by the power plant owners, it logically could be considered an extension of the plant itself. In such cases, the CO_2 pipelines could be eligible for regulated returns on the invested capital and their costs could be recovered by utilities in electricity rates. Alternatively such a CO_2 pipeline could be owned by third parties and considered a non-plant asset providing a transportation service for a fee. In the latter case, the costs could still be recovered by the utility in its rates as an operating cost.

Two complications arise with respect to pipeline cost recovery. First, because utility regulation varies from state to state (e.g., some states allow for competition in electricity generation, others do not),[54] differences among states in the economic regulation of CO_2 pipelines could create economic inefficiencies and affect the attractiveness of CO_2 pipelines for capital investment. Second, if CO_2 transportation infrastructure is intended to evolve from shorter, stand-alone, *intra*state pipelines into a network of interconnected *inter*state pipelines, pipeline operators wishing to link CO_2 pipelines across state lines may face a regulatory environment of daunting complexity. Without a coherent system of economic regulation for CO_2 pipelines, whether as a commodity, pollutant, or some other classification, developers of interstate CO_2 pipelines may need to negotiate or litigate repeatedly issues such as siting, pipeline access, terms of service, and rate "pancaking" (the accumulation of transportation charges assessed by contiguous pipeline operators along a particular transportation route). It is just these kinds of issues which have complicated and impeded the integration of individual utilityelectric transmission systems into larger regional transmission networks.[55]

CO_2 Pipeline Incentives

Oil industry representatives frequently point to EOR as offering a market-based model for profitable CO_2 transportation via pipeline. It should be noted, however, that much of the existing CO_2 pipeline network in the United Statesfor EOR has been established with the benefit of federal tax incentives. Although current federal tax law provides no special or targeted tax benefits specifically to CO_2 pipelines, investments in CO_2 pipelines *do* benefit from tax provisions targeted for EOR. They also benefit from accelerated depreciation rules, which apply generally to any capital investment including petroleum and natural gas (non-CO_2) pipelines. For example, the Internal Revenue Code provides for a 15% income tax credit for the costs of recovering domestic oil by one of nine qualified EOR methods, including CO_2 injection (I.R.C. § 43).[56] Also, extraction of naturally occurring CO_2 may qualify for percentage depletion allowance under

I.R.C. § 613(b)(7). Prior federal law, both tax and nontax, also provided various types of incentives for EOR which stimulated investment in CO_2 pipelines. In particular, oil produced from EOR projects was exempt from oil price controls in the 1970s. Development of CO_2 pipeline infrastructure in the 1980s benefitted from tax advantages to EOR oil under the crude oil windfall profits tax law, which was in effect from March 1980 to August 1988.

Although there were never incentives explicitly for CO_2 pipelines under federal tax and price control regulation in the 1970s and 1980s, it is clear that CO_2 pipeline infrastructure development benefitted from these regulations. In a CCS environment where some captured CO_2 is a valuable commodity, but the remainder is not, establishing similar regulatory incentives for CO_2 pipelines becomes complex. One initial proposal in S. 2149 would allow seven-year accelerated depreciation for qualifying CO_2 pipelines constructed after enactment (Sec. 4). As debate continues about the economics of CO_2 capture and sequestration generally, and how the federal government can encourage CCS infrastructure investment, Congress may seek to understand the implications of CCS incentives specifically on CO_2 pipeline development.

Cost Implications for Network Development

In light of the overall costs associated with CO_2 pipelines, including the uncertainty about future materials costs and cost recovery, some analysts anticipate that a CO_2 network for CCS will begin with shorter pipelines from CO_2 sources located close to sequestration sites. Larger CO_2 trunk lines are expected to emerge to capture substantial scale economies in long-distance pipeline transportation. According to the 2007 MIT report, "it is anticipated that the first CCS projects will involve plants that are very close to a sequestration site or an existing CO_2 pipeline. As the number of projects grow, regional pipeline networks will likely evolve."[57] It is debatable, however, whether piecemeal growth of a CO_2 pipeline network in this way, presumably by individual facility operators seeking to minimize their own costs, would ultimately yield an economically efficient and publically acceptable CO_2 pipeline network for CCS. Weaknesses and failures in the North American electric power transmission grid, which was developed in this manner, may be one example of how piecemeal, uncoordinated network development may fail to satisfy key economic and operating objectives.

As an alternative to piecemeal CO_2 pipeline development, some analysts suggest that it may be more cost effective in the long run to build large trunk pipelines when the first sites with CO_2 capture come on line with the expectation that subsequent users could fill the spare capacity in the trunk line. In addition to lower per-unit transport costs for CO_2, such an arrangement would smooth out potentially intermittent CO_2 flows from individual capture sites (especially discontinuously operated power plants), provide a greater buffer for overall CO_2 supply fluctuations, and generally allow for more operational flexibility in the system.[58] Planning and financing such a CO_2 trunk line system would present its own challenges, however. As another analysis points out, "implementation of a 'backbone' transport structure may facilitate access to large remote storage reservoirs, but infrastructure of this kind will require large initial upfront investment decisions."[59] How a CO_2 network for CCS would be configured, and who would configure it, may be issues for Congress.[60]

CO_2 Pipeline Safety

CO_2 occurs naturally in the atmosphere, and is produced by the human body during ordinary respiration, so it is commonly perceived by the general public to be a relatively harmless gas. However, at concentrations above 10% by volume, CO_2 may cause adverse health effects and at concentrations above 25% poses a significant asphyxiation hazard. Because CO_2 is colorless, odorless, and heavier than air, an uncontrolled release may accumulate and remain undetected near the ground in low-lying outdoor areas, and in confined spaces such as caverns, tunnels, and basements.[61] Exposure to CO_2 gas, as for other asphyxiates, may cause rapid "circulatory insufficiency," coma, and death.[62] Such an event occurred in 1986 in Cameroon, when a cloud of naturally-occurring CO_2 spontaneously released from Lake Nyos killed 1,800 people in nearby villages.[63]

The Secretary of Transportation has primary authority to regulate interstate CO_2 pipeline safety under the Hazardous Liquid Pipeline Act of 1979 as amended (49 U.S.C. § 601). Under the act, the Department of Transportation (DOT) regulates the design, construction, operation and maintenance, and spill response planning for CO_2 pipelines (49 C.F.R. § 190, 195-199). The DOT administers pipeline regulations through the Office of Pipeline Safety (OPS) within the Pipelines and Hazardous Materials Safety Administration (PHMSA).[64] Although CO_2 is listed as a Class 2.2 (non-flammable gas) hazardous material under DOT regulations (49 C.F.R. § 172.101), the agency applies nearly the same safety requirements to CO_2 pipelines as it does to pipelines carrying hazardous liquids such as crude oil, gasoline, and anhydrous ammonia (49 C.F.R. § 195).

To date, CO_2 pipelines in the United States have experienced few serious accidents. According to OPS statistics, there were 12 leaks from CO_2 pipelines reported from 1986 through 2006 — none resulting in injuries to people. By contrast, there were 5,610 accidents causing 107 fatalities and 520 injuries related to natural gas and hazardous liquids (excluding CO_2) pipelines during the same period.[65] It is difficult to draw firm conclusions from these accident data, because CO_2 pipelines account for less than 1% of total natural gas and hazardous liquids pipelines, and CO_2 pipelines currently run primarily through remote areas. Based on the limited sample of CO_2 incidents, analysts conclude that, mile-for-mile, CO_2 pipelines appear to be safer than the other types of pipeline regulated by OPS.[66] Additional measures, such as adding gas odorants to CO_2 to aid in leak detection, may further mitigate CO_2 pipeline hazards. Nonetheless, as the number of CO_2 pipelines expands, analysts suggest that "statistically, the number of incidents involving CO_2 should be similar to those for natural gas transmission."[67] If the nation's CO_2 pipeline network expands significantly to support CCS, and if this expansion includes more pipelines near populated areas, more CO_2 pipeline accidents are likely in the future.[68]

Criminal and Civil Liability

There are no special provisions in U.S. law protecting the CO_2 pipeline industry from criminal or civil liability. In January 2003, the Justice Department announced over $100 million in civil and criminal penalties against Olympic Pipeline and Shell Pipeline resolving claims from a fatal gasoline pipeline fire in Bellingham, WA, in 1999.[69] In March 2003, emphasizing the environmental aspects of homeland security, Attorney General John Ashcroft reportedly announced a crackdown on companies failing to protect against possible terrorist attacks on storage tanks, transportation networks, industrial plants, and pipelines.[70]

Even if no federal or state regulations are violated, CO_2 pipeline operators could still face civil liability for personal injury or wrongful death in the event of an accident. In the Bellingham accident, the pipeline owner and associated defendants reportedly agreed to pay a $75 million settlement to the families of two children killed in the accident.[71] In 2002, El Paso Corporation settled wrongful death and personal injury lawsuits stemming from a natural gas pipeline explosion near Carlsbad, NM, which killed 12 campers.[72] Although the terms of those settlements were not disclosed, two additional lawsuits sought a total of $171 million in damages.[73] The MIT study concluded that operational liability for CO_2 pipelines, as part of an integrated CCS infrastructure, "can be managed within the framework that has been successfully used for decades by the oil and gas industries."[74] Nonetheless, as CCS policy evolves, Congress may seek to ensure that liability provisions for CO_2 pipelines are adequate and consistent with liability provisions in place for other CO_2 infrastructure.

Other Issues

Inaddition to the issues discussed above, additional policy issues related to CO_2 pipelines may arise as CCS policy evolves. These may include addressing technical transportation problems related to the presence of other pollutants, such as sulfuric and carbonic acid, in CO_2 pipelines. Some have also suggested the use or conversion of existing non-CO_2 pipelines, such as natural gas pipelines, to transport CO_2.[75]

Coordination of U.S. CO_2 pipeline policies with Canada, with whom the United States shares its existing pipeline infrastructure, may also become a consideration. Finally, the potential impacts of CO_2 pipeline development overseas on the global availability of construction skills and materials may arise as a key factor in CCS economics and implementation.

CONCLUSION

Policy debate about the mitigation of climate change through some scheme of carbon capture and sequestration is expanding quickly. To date, debate among legislators has been focused mostly on CO_2 sources and storage sites, but CO_2 pipelines are a vital connection between the two. Although CO_2 transportation by pipeline is in some respects a mature technology, there are many important unanswered questions about the socially optimal configuration, regulation, and costs of a CO_2 pipeline network for CCS. Furthermore, because CO_2 pipelines for EOR are already in use today, policy decisions affecting CO_2 pipelines take on an urgency that is, perhaps, unrecognized by many. It appears, for example, that federal classification of CO_2 as both a commodity (by the BLM) and as a pollutant (by the EPA) potentially could create an immediate conflict which mayneed to be addressed not only for the sake of future CCS implementation, but also to ensure consistency between future CCS and today's CO_2 pipeline operations.

In addition to these issues, Congress may examine how CO_2 pipelines fit into the nation's overall strategies for energy supply and environmental protection. The need for CO_2 pipelines ultimately derives from the nation's consumption of fossil fuels. Policies affecting the latter, such as energy conservation, and the development of new renewable, nuclear, or hydrogen energy resources, could substantially affect the need for and configuration of CO_2 pipelines. If policy makers encourage continued consumption of fossil fuels under CCS, then the need to foster the other energy options may be diminished — and vice versa. Thus decisions about CO_2 pipeline infrastructure could have consequences for a broader array of energy and environmental policies.

End Notes

[1] This chapter does not explore the underlying science of climate change, nor the question of whether action is justified. See CRS Report RL33849, *Climate Change: Science and Policy Implications*, by Jane A. Leggett.

[2] For more information on congressional activities related to global warming, see CRS Report RL31931, *Climate Change: Federal Laws and Policies Related to Greenhouse Gas Reductions*, by Brent D. Yacobucci and Larry Parker; and CRS Report RL34067, *Climate Change Legislation in the 110th Congress*, by Jonathan L. Ramseur and Brent D. Yacobucci.

[3] This chapter does not address indirect sequestration, wherein CO_2 is stored in soils, oceans, or plants through natural processes. For information on the latter, see CRS Report RL31432, *Carbon Sequestration in Forests*, by Ross W. Gorte.

[4] John Douglas, "Expanding Options for CO_2 Storage," *EPRI Journal,* Electric Power Research Institute (Spring 2007): 24.

[5] More detailed information is available in CRS Report RL33801, *Direct Carbon Sequestration: Capturing and Storing CO_2*, by Peter Folger.

[6] Intergovernmental Panel on Climate Change, Special Report: *Carbon Dioxide Capture and Storage, 2005* (2005): 22-23. (Hereafter referred to as IPCC 2005.)

[7] H. J. Herzog and D. Golumb, "Carbon Capture and Storage from Fossil Fuel Use," in C.J. Cleveland (ed.), *Encyclopedia of Energy* (New York, NY: Elsevier Science, Inc., 2004): 277-287.

[8] IPCC 2005: 26.

[9] IPCC 2005: 181.

[10] CO_2 used in EOR enhances oil production by re-pressurizing geological formations and reducing oil viscosity, thereby increasing oil movement to the surface. CO2 is used industrially as a chemical feedstock, to carbonate beverages, for refrigeration and food processing, to treat water, and for other uses.

[11] IPCC 2005: 31.

[12] Sedimentary basins are large depressions in the Earth's surface filled with sediments and fluids.

[13] Kinder Morgan CO_2 Company, "Canyon Reef Carriers Pipeline (CRC)," web page (2007). [http://www.kindermorgan.com/business/co2/transport_canyon_reef.cfm]

[14] U.S. Dept. of Transportation, National Pipeline Mapping System database (June 2005). [https://www.npms.phmsa.dot.gov]

[15] Bureau of Transportation Statistics (BTS), National Transportation Statistics 2005 (Dec. 2005), Table 1-10. In this chapter *oil* includes petroleum and other hazardous liquids such as gasoline, jet fuel, diesel fuel, and propane, unless otherwise noted.

[16] R.T. Dahowski, J.J. Dooley, C.L. Davidson, S. Bachu, N. Gupta, and J. Gale, "A North American CO_2 Storage Supply Curve: Key Findings and Implications for the Cost of CCS Deployment," *Proceedings of the Fourth Annual Conference on Carbon Capture and Sequestration* (Alexandria, VA: May 2-5, 2005). The study addresses CO_2 capture at 2,082 North American facilities including power plants, natural gas processing plants, refineries, cement kilns, and other industrial plants.

[17] Jennie C. Stevens and Bob Van Der Zwaan, "The Case for Carbon Capture and Storage," *Issues in Science and Technology*, vol. XXII, no. 1 (Fall 2005): 69-76. (See page 15 of this chapter for a discussion of safety issues.)

[18] John Deutch, Ernest J. Moniz, et al., *The Future of Coal.* (Cambridge, MA: Massachusetts Institute of Technology: 2007): 58. (Hereafter referred to as MIT 2007.)

[19] MIT 2007: 58.

[20] Eric Williams, Nora Greenglass, and Rebecca Ryals, "Carbon Capture, Pipeline and Storage: A Viable Option for North Carolina Utilities?" Working paper prepared by the Nicholas Institute for Environmental Policy Solutions and The Center on Global Change, Duke University (Durham, NC: March 8, 2007): 4.

[21] MIT 2007: 58.

[22] General Accounting Office (now Government Accountability Office), *Surface Transportation: Issues Associated With Pipeline Regulation by the Surface Transportation Board*, RCED-98-99 (Washington, DC: April 21, 1998):3; and 49 U.S.C. § 155.

[23] The STB is the successor agency to the Interstate Commerce Commission (ICC) under the Interstate Commerce Commission Termination Act of 1995 (P.L. 104-88).

[24] For a more comprehensive discussion of CO_2 pipeline regulatory jurisdiction, see CRS Report RL34307, *Regulation of Carbon Dioxide (CO_2) Sequestration Pipelines: Jurisdictional Issues*, by Adam Vann and Paul W. Parfomak.

[25] 15 U.S.C. 717f(c).

[26] 49 App. U.S.C.§1.

[27] Section 1801 of the Energy Policy Act of 1992 directed FERC to "promulgate regulations establishing a simplified and generally applicable ratemaking methodology" for oil pipeline transportation.

[28] 49 U.S.C. § 1-501(a)(1)(c).

[29] *Cortez Pipeline Company,* 7 FERC ¶ 61,024 (1979).

[30] Id.

[31] Cortez Pipeline Company — Petition for Declaratory Order — Commission Jurisdiction Over Transportation of Carbon Dioxide by Pipeline, 46 Fed. Reg. 18805 (March 26, 1981).

[32] Now known as the Government Accountability Office.

[33] Surface Transportation Board (STB), Personal communication, (December 2007). The STB Office of Governmental and Public Affairs informed CRS that the board recognizes the conflict between this GAO report and the ICC decision (as well as the wording of 49 C.F.R. § 15301 governing STB jurisdiction over pipelines other than those transporting "water, gas or oil"). However the office did not want to state an opinion as to the current extent of STB jurisdiction over CO_2 pipelines and suggested that the STB would likely not act to resolve this conflict unless a CO_2 pipeline dispute comes before it.

[34] Chevron U.S.A. v. Nat. Res. Def. Council, 467 U.S. 837, at 863-64 (1984).

[35] Beard Company 2000 annual report (10-k) filed with the U.S. Securities and Exchange Commission states that the company (with other plaintiffs) filed a lawsuit in 1996 against CO2 pipeline owner Shell Oil Company and other defendants alleging, among other things, that the defendants "controlled and depressed the price of CO2" from a field partially owned by Beard and "reduc[ed] the delivered price of CO2 while ... simultaneously inflating the cost of transportation." [http://www.secinfo.com/dRxzp.424.htm#1fmr]

[36] 18 C.F.R. § 157.

[37] National Commission on Energy Policy, *Siting Critical Energy Infrastructure: An Overview of Needs and Challenges*. (Washington, DC: June 2006): 9. (Hereafter referred to as NCEP 2006.)

[38] Partha S. Chaudhuri, Michael Murphy, and Robert E. Burns, "Commissioner Primer: Carbon Dioxide Capture and Storage" (National Regulatory Research Institute, Ohio State Univ., Columbus, OH: Mar. 2006): 17.

[39] U.S. Dept. of the Interior, Bureau of Land Management, *Environmental Assessment for Anadarko E&P Company L.P. Monell CO2 Pipeline Project*, EA #WY-040-03-035 (Feb. 2003): 71.

[40] Chaudhuri et al: 17.

[41] NCEP 2006: 9.

[42] IOGCC 2005: 41.

[43] S.M. Frailey, R.J. Finlay, and T.S. Hickman, "CO_2 Sequestration: Storage Capacity Guideline Needed," *Oil & Gas Journal* (Aug. 14, 2006): 44.

[44] *Massachusetts v. EPA*; at [http://www.supremecourtus.gov/opinions/06pdf/05-1120.pdf]. For further information see CRS Report RL33776, *Clean Air Issues in the 110th Congress: Climate Change, Air Quality Standards, and Oversight*, by James E. McCarthy.

[45] U.S. Environmental Protection Agency, memorandum (July 5, 2006). Available at [http://www.epa.gov/OGWDW/uic/pdfs/memo_wells_sequestration_7-5-06.pdf].

[46] MIT 2007: xi.

[47] Eric Williams et al. (2007): 20.

[48] N. Parker, "Using Natural Gas Transmission Pipeline Costs to Estimate Hydrogen Pipeline Costs," UCD-ITS-RR-04-35, Inst. of Transportation Studies, Univ. of California (Davis, CA: 2004): 1. [http://hydrogen.its.ucdavis.edu/people/ncparker/papers/pipelines]; see, also, G. Heddle, H. Herzog, and M. Klett, "The Economics of CO2 Storage," MIT LFEE 2003-003 RP (Laboratory for Energy and the Environment, MIT, Cambridge, MA: Aug. 2003). [http://lfee.mit.edu/public/LFEE_2003-003_RP.pdf]

[49] MIT 2007: 58.

[50] N. Parker (2004): Fig. 23.

[51] IPCC 2005: 27.

[52] See CRS Report RL32333, *Steel: Price and Policy Issues*, by Stephen Cooney.

[53] Michael Cowden, "A Profusion of New Pipeline Projects and Profits... for Now," *American Metal Market* (January 2008): 18; Global Insight, *Steel Industry Review* (2nd Qtr. 2006), tabs. 1.11-1.12; and *American Metal Market*, "West Sees More Steel Plate But Prices Holding Ground" (Aug. 31, 2006).

[54] In market-based states, cost recovery may affect electricity markets.

[55] For further information of electric transmission regulation, see CRS Report RL33875, *Electric Transmission: Approaches for Energizing a Sagging Industry*, by Amy Abel.

[56] Unfortunately for EOR investors, while this tax credit is part of current federal tax law, its phaseout provisions mean that presently it is not available — the credit is zero — due to high crude oil prices.

[57] Ibid., MIT. (2007): 59.

[58] John Gale and John Davidson, "Transmission of CO2 — Safety and Economic Considerations," *Energy,* Vol. 29, Nos. 9-10 (July-August 2004): 1326.

[59] IPCC 2005: 190.

[60] For further discussion see CRS Report RL34316, *Pipelines for Carbon Dioxide (CO2) Control: Network Needs and Cost Uncertainties*, by Paul W. Parfomak and Peter Folger.

[61] J. Barrie, K. Brown, P.R. Hatcher, and H.U. Schellhase, "Carbon Dioxide Pipelines: A Preliminary Review of Design and Risks," Proceedings of the 7th International Conference on Greenhouse Gas Control Technologies (Vancouver, Canada: Sept. 5-9, 2004): 2.

[62] Airco, Inc., "Carbon Dioxide Gas," *Material Safety Data Sheet* (Aug. 4, 1989). [http://www2.siri.org/msds/f2/byd/bydjl.html]

[63] Kevin Krajick, "Defusing Africa's Killer Lakes," *Smithsonian*, v. 34, n. 6. (2003): 46 —55.

[64] PHMSA succeeds the Research and Special Programs Administration (RSPA), reorganized under P.L. 108-246, which was signed by the President on Nov. 30, 2004.

[65] Office of Pipeline Safety (OPS), "Distribution, Transmission, and Liquid Accident and Incident Data," (2007). OPS has not yet released 2007 incident statistics. Data files available at [http://ops.dot.gov/stats/IA98.htm].

[66] John Gale and John Davidson. (2004): 1322.

[67] Barrie et al. (2004): 2.

[68] Gale and Davidson (2004): 1321.

[69] "Shell, Olympic Socked for Pipeline Accident," *Energy Daily* (Jan. 22, 2003).

[70] John Heilprin, "Ashcroft Promises Increased Enforcement of Environmental Laws for Homeland Security," *Assoc. Press,* Washington dateline (Mar. 11, 2003).

[71] Business Editors, "Olympic Pipe Line, Others Pay Out Record $75 Million in Pipeline Explosion Wrongful Death Settlement," *Business Wire* (April 10, 2002).

[72] National Transportation Safety Board, *Pipeline Accident Report,* PAR-03-01. (Feb. 11, 2003).

[73] El Paso Corp., *Quarterly Report Pursuant to Section 13 or 15(d) of the Securities Exchange Act of 1934,* Form 10-Q, Period ending June 30, 2002. (Houston, TX: 2002). The impact of these lawsuits on the company's business is unclear, however; the report states that "our costs and legal exposure ... will be fully covered by insurance."

[74] MIT 2007: 58.

[75] An example is the Gwinville, MS-Lake St. John, LA natural gas pipeline purchased by Denbury Resources, Inc. in 2006 and converted to CO2 transportation for EOR in 2007.

In: Pipelines for Carbon Sequestration: Background ... ISBN: 978-1-60741-383-7
Editors: Elvira S. Hoffmann

Chapter 2

PIPELINES FOR CARBON DIOXIDE (CO_2) CONTROL: NETWORK NEEDS AND COST UNCERTAINTIES

Paul W. Parformak and Peter Folger
Resources, Science, and Industry Division

SUMMARY

Congress is considering policies promoting the capture and sequestration of carbon dioxide (CO_2) from sources such as electric power plants. Carbon capture and sequestration (CCS) is a process involving a CO_2 source facility, a long-term CO_2 sequestration site, and CO_2 pipelines. There is an increasing perception in Congress that a national CCS program could require the construction of a substantial network of interstate CO_2 pipelines. However, divergent views on CO_2 pipeline requirements introduce significant uncertainty into overall CCS cost estimates and may complicate the federal role, if any, in CO_2 pipeline development. S. 2144 and S. 2191 would require the Secretary of Energy to study the feasibility of constructing and operating such a network of pipelines. S. 2323 would require carbon sequestration projects to evaluate the most cost-efficient ways to integrate CO_2 sequestration, capture, and transportation. P.L. 110-140, signed by President Bush on December 19, 2007, requires the Secretary of the Interior to recommend legislation to clarify the issuance of CO_2 pipeline rights-of-way on public land.

The cost of CO_2 transportation is a function of pipeline length and other factors. This chapter examines key uncertainties in CO_2 pipeline requirements for CCS by contrasting hypothetical pipeline scenarios for 11 major coal-fired power plants in the Midwest Regional Carbon Sequestration Partnership region. The scenarios illustrate how different assumptions about sequestration site viability can lead to a 20-fold difference in CO_2 pipeline lengths, and, therefore, similarly large differences in capital costs. From the perspective of individual power plants, or other CO_2 sources, variable costs for CO_2 pipelines may have significant ramifications. If CO_2 pipeline costs for specific regions reach tens, or even hundreds, of millions of dollars per plant, then power companies may have difficulty securing the capital financing or regulatory approval needed to construct or retrofit fossil fuel-powered plants in these regions. High CO_2 transportation costs also could increase electricity prices in "sequestration-poor" regions relative to regions able to sequester CO_2 more locally.

As CO_2 pipelines get longer, the state-by-state siting approval process may become complex and protracted, and may face public opposition. Because CO_2 pipeline requirements in a CCS scheme are driven by the relative locations of CO_2 sources and sequestration sites, identification and validation of such sites must explicitly account for CO_2 pipeline costs if the economics of those sites are to be fully understood. Since transporting CO_2 to distant locations can impose significant additional costs to a facility's carbon control infrastructure, facility owners may seek regulatory approval for as many sequestration sites as possible and near to as many facilities as possible. If CCS moves to widespread implementation, government agencies and private companies may face challenges in identifying, permitting, developing, and monitoring the large number of localized sequestration reservoirs that may be proposed. However, even as viable sequestration reservoirs are being identified, it is unclear which CO_2 source facilities will have access to them, under what time frame, and under what conditions. Given the potential size of a national CO_2 pipelines network, many billions of dollars of capital investment may be affected by policy decisions made today.

INTRODUCTION

Congress is considering policies to reduce U.S. emissions of greenhouse gases. Prominent among these policies are those promoting the capture and direct sequestration of carbon dioxide (CO_2) from manmade sources such as electric power plants and manufacturing facilities. Carbon capture and sequestration is of great interest because potentially large amounts of CO_2 produced by the industrial burning of fossil fuels could be sequestered. Although they are still under development, carbon capture technologies may be able to remove up to 95% of CO_2 emitted from an electric power plant or other industrial source.

Carbon capture and sequestration (CCS) is a three-part process involving a CO_2 source facility, a long-term CO_2 sequestration site, and an intermediate mode of CO_2 transportation — typically pipelines. Some studies have been optimistic about pipeline requirements for CO_2 sequestration. They conclude that the pipeline technology is mature, and that most major CO_2 sources in the United States are, or will be, located near likely sequestration sites, so that large investments in CO_2 pipeline infrastructure will probably not be needed.[1] Other studies express greater uncertainty about the required size and configuration of CCS pipeline networks.[2] A handful of regionally-focused studies have concluded that CO_2 pipeline requirements for CO_2 sources could be substantial, and thus present a greater challenge for CCS than is commonly presumed, at least in parts of the United States.[3]

Divergent views on CO_2 pipeline requirements introduce significant uncertainty into overall CCS cost estimates and may complicate the federal role, if any, in CO_2 pipeline regulation. They are also a concern because uncertainty about CO_2 pipeline requirements may impede near-term capital investment in electricity generation, with important implications for power plant owners seeking to reduce their CO_2 emissions.

In the 110th Congress, there has been considerable debate on the capture and sequestration aspects of carbon sequestration, while there has been relatively less focus on transportation. Nonetheless, there is an increasing perception in Congress that a national CCS program could require the construction of a substantial network of interstate CO_2 pipelines. The Carbon Dioxide Pipeline Study Act of 2007 (S. 2144), introduced by Senator Norm Coleman and nine cosponsors on October 4, 2007, would require the Secretary of Energy to study the feasibility of constructing and operating such a network of CO_2 pipelines. The America's Climate Security Act of 2007 (S. 2191), introduced by Senator Joseph Lieberman and nine cosponsors on October 18, 2007, and reported out of the Senate Environment and Public Works Committee in amended form on December 5, 2007, contains similar provisions (Sec. 8003). The Carbon Capture and Storage Technology Act of 2007 (S. 2323), introduced by Senator John Kerry and one cosponsor on November 7, 2007, would require carbon sequestration projects authorized by the act to evaluate the most cost-efficient ways to integrate CO_2 sequestration, capture, and transportation (Sec. 3(b)(5)). The Energy

Independence and Security Act of 2007 (P.L. 110-140), signed by President Bush, as amended, on December 19, 2007, requires the Secretary of the Interior to recommend legislation to clarify the appropriate framework for issuing CO_2 pipeline rights-of-way on public land (Sec. 714(7)).

This chapter examines key uncertainties in CO_2 pipeline requirements for CCS by contrasting hypothetical pipeline scenarios in one region of the United States. The report summarizes the key factors influencing CO_2 pipeline configuration for major power plants in the region, and illustrates how the viability of different sequestration sites may lead to enormous differences in pipeline costs. Power plants, particularly coal-fired plants, are the most likely initial candidates for CCS because they are predominantly large, single-point sources, and they contribute approximately one-third of U.S. CO_2 emissions from fossil fuels. The report discusses the implications of uncertain CO_2 pipeline requirements for CCS as they relate to evolving federal policies for carbon control.

Scenarios for CO_2 Pipeline Development

Under a national CCS policy, a key question is how to establish a CO_2 pipeline network atthe lowest social and economic cost given the current locations of existing CO_2 source facilities and the locations of future sequestration sites. On its face, this may appear to be a straightforward analytic problem of the type regularly addressed in other network industries. The oil and gas industry, among others, employs myriad analytic techniques to identify and optimize potential routes for new fuel pipelines.[4] In the context of CCS, however, predicting pipeline routes is more challenging because there is considerable uncertainty about the suitability of geological formations to sequester captured CO_2 and the proximity of suitable formations to specific sources of CO_2. One recent analysis, for example, concluded that 77% of the total annual CO_2 captured from the major North American sources could be stored in reservoirs directly underlying these sources, and that an additional 18% could be stored within 100 miles of the original sources.[5] Other analysts suggest that captured CO_2 may need to be sequestered, at least initially, in more centralized reservoirs to reduce potential risks associated with CO_2 leaks.[6] They suggest that, given current uncertainty about the suitability of various on-site geological formations for long-term CO_2 sequestration, certain specific types of formations (e.g., saline aquifers) may be preferred as CO_2 repositories because they have adequate capacity and are most likely to retain sequestered CO_2 indefinitely.

The Department of Energy estimates that the United States has enough capacity to store CO_2 for tens to hundreds of years.[7] However, the large-scale CO_2 experiments needed to acquire detailed data about potential sequestration reservoirs have only just begun. Given current uncertainty about potential sequestration sites, policy discussions about CCS envision various possible scenarios for the development of a CO_2 pipeline network. If CO_2 can be sequestered near where it is produced then CO_2 pipelines might evolve in a decentralized way, with individual facilities developing direct pipeline connections to nearby sequestration sites largely independent of other companies' pipelines. The resulting network might then consist of many relatively short and unconnected pipelines with a small number of longer pipelines for facilities with no sequestration sites nearby. Alternatively, if only very large, centralized sequestration sites are permitted, the result might be a network of interconnected long distance pipelines, perhaps including high-capacity trunk lines serving a multitude of feeder pipelines from individual facilities. A third scenario envisions CO_2 sequestration, at least initially, at active oil fields where injection of CO_2 may be profitably employed for enhanced oil recovery (EOR). Indeed, a CO_2 pipeline network already exists for EOR purposes in the southwestern United States, although it is limited in geographic reach. Whether CCS policies ultimately lead to one or more of these scenarios remains to be seen; however, the configuration of the resulting CO_2 pipeline network, and its associated costs, may have a significant bearing on which CCS policies best serve the public interest.

Hypothetical CO_2 Pipelines in the Midwest

Infrastructure requirements and policy implications related to CO_2 pipelines become clearer when considering what actual pipeline projects might look like. This section outlines contrasting scenarios for hypothetical CO_2 pipeline development in the region covered by the Midwest Regional Carbon Sequestration Partnership (MRCSP). The MRCSP is one of seven regional partnerships of state agencies, universities, private companies, and non-governmental organizations established by the Department of Energy to assess CCS approaches. The MRCSP serves as a good illustration of CO_2 pipeline issues because it has a varied mix of CO_2 sources and potential geologic sequestration sites, and because geologists have completed a number of focused studies relevant to CCS in this region.

Sequestration in the Rose Run Formation

The MRCSP has identified key CO_2 sources and geologic formations potentially suitable for carbon sequestration within its seven-state region encompassing northeast Indiana, Kentucky, Maryland, Michigan, Ohio, Pennsylvania, and West Virginia. **Figure 1** shows the locations of 11 of the largest CO_2 sources located in the MRSCP region — all coal-fired electric power plants emitting over 9 million metric tons of CO_2 annually.[8] There are numerous other CO_2 sources in this region, including many other power plants and large industrial facilities, but the 11 power plants in this analysis include the very largest in terms of annual CO_2 emissions.

Figure 1 also shows the locations of the Rose Run sandstone, a deep saline formation identified by the MRCSP as a potential carbon sequestration site.[9] As the figure shows, the plants all lie above or near to this formation, so suitable CO_2 injection sites presumably could be located very near to each of these plants. If the Rose Run formation proves to be viable for large-scale CO_2 sequestration, then some plants may be able to inject CO_2 directly below their facilities, and CCS pipeline requirements for some of the other 11 power plants could be small. If this were the case, then the CCS CO_2 pipeline network for the 11 plants might appear as shown in **Figure 2**.

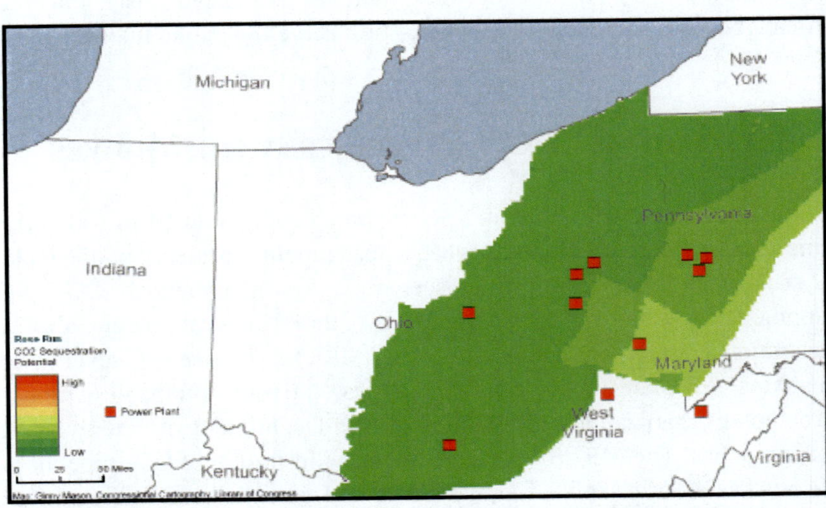

Source: MRCSP. Geologic data for NY are not provided by the MRCSP.

Figure 1. Major Power Plants and the Rose Run Formation

The hypothetical pipeline layout in **Figure 2** assumes that a 25-mile diameter, non-overlapping reserve area is needed for each plant's sequestration site and that any location within the Rose Run formation is viable for sequestration. **Figure 2** also assumes that each power plant is either located at or is connected to the center of its respective sequestration field by a large trunk pipeline built along existing rights of way and capable of carrying its peak CO_2 output. Smaller pipelines branching from the centrally-located plant or from the trunk line distribute the CO_2 to multiple injection wells in the sequestration site. These smaller pipelines are not considered in detail in this chapter.

Figure 2 shows that the longest trunk pipeline for CO_2transportation is 32 miles long, and the average pipeline is approximately 11 miles long. According to models developed at Carnegie Mellon University (CMU), the capital costs to construct an 11-mile pipeline in the Midwestern United States with a capacity of 10 million tons of CO_2 annually would be approximately $6 million. The levelized cost would be approximately $0.10 per ton of transported CO_2, including costs for operation (e.g., compression) and maintenance.[10]

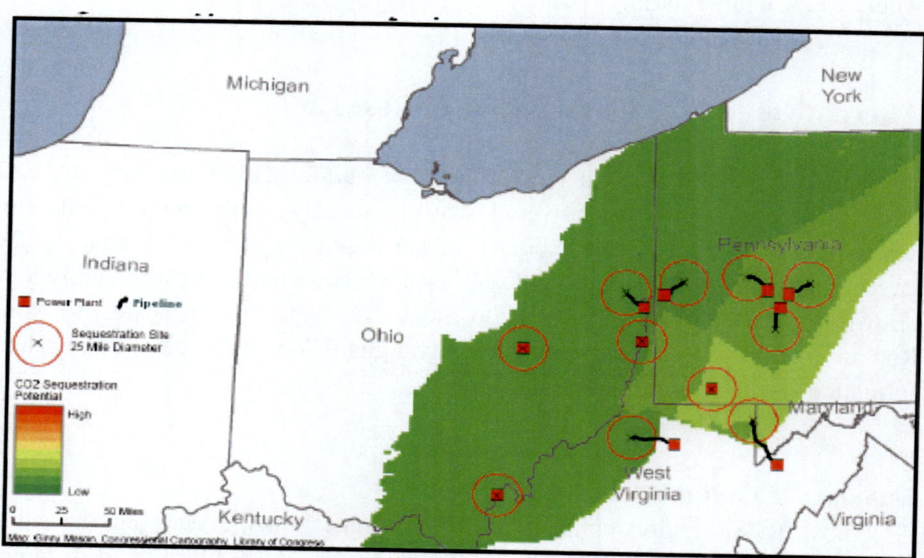

Source: MRCSP, CRS. Geologic data for NY are not provided by the MRCSP

Figure 2. Hypothetical CO_2 Pipelines to the Rose Run Formation

Potential Barriers to Rose Run Sequestration

Although the Rose Run formation is identified by the MRCSP as a major potential sequestration site, it has characteristics which may ultimately limit its viability for large-scale CO_2 sequestration. The most important of these is overall sequestration capacity. Because the Rose Run formation has low to moderate permeability and thickness, geologic models show that it is unlikely all of the CO_2 emitted in the Rose Run region can be efficiently sequestered in the Rose Run formation.[11] The Rose Run formation is also relatively fractured.[12] Geologists have concluded that injecting pressurized CO_2 into the Rose Run formation potentially could induce minor earthquakes along certain preexisting (but undetected) faults in otherwise seismically stable areas.[13] Faults and fractures can, in some cases, provide additional sequestration capacity and be beneficial for sequestration. But faults or fractures can also be permeable conduits for leakage and "can be a significant pathway for the loss of sequestered CO_2."[14] While studies are not yet available to establish the validity of any of these concerns, future research mayconclude that significant parts of the Rose Run formation would be unsuitable for large scale, permanent CO_2 sequestration.[15]

Alternatives to CO_2 Sequestration in Rose Run

The CO_2 sequestration capacity of the Rose Run formation may turn out to be too limited because of its of overall size or integrity. If the policy goal is to sequester CO_2 from all major sources in the region, then at least some of the largest power plants in the MRCSP will need to sequester their carbon emissions elsewhere. The alternative sites for potential CO_2 sequestration nearest to Rose Run are unmineable coal beds, oil and natural gas fields, and another large saline formation — the Mount Simon sandstone.

Unmineable Coal Beds

The MRCSP region contains unmineable coal beds underlying the same general geographic footprint as the Rose Run formation, but located at different depths underground. Studies suggest that such coal beds may be suitable for sequestration. In some cases injected CO_2 could replace methane trapped in the coal seam, increasing natural gas available for extraction wells in a process similar to EOR known as enhanced coal-bed methane recovery. However, the potential capacity for storing CO_2 in regional coal beds is only about 5%

compared to the Rose Run sandstone, and the practicability of storing CO_2 in coal seams is virtually untested.[16] In addition, removing groundwater from coal seams prior to CO_2 injection may create environmental problems related to water disposal, and some studies indicate that coal swelling associated with CO_2 injection may curtail the permeability of the coal seam, limiting its overall capacity to store CO_2.[17]

Oil and Natural Gas Fields

The MRCSP region includes a number of oil and natural gas fields which may offer opportunities for CO_2 sequestration. The region also includes a number of natural gas storage reservoirs, both natural and manmade, which suggest that CO_2 could be similarly stored. However, according to the MRCSP, the ten largest oil and gas fields in the region have an average CO_2 sequestration potential of only 251 million tons.[18] By comparison, the 30-year CO_2 output of the 11 plants in this analysis would range from 270 to 491 million tons at current emission levels. The oil and gas fields in the MRCSP region, therefore, even if they could achieve their stated sequestration potential, may not individually have sufficient capacity to sequester CO_2 from one of the 11 power plants in this analysis operating with current emissions over a 30-year period. Multiple fields possibly could be used by individual power plants to achieve adequate long-term sequestration, but this would require multiple pipeline networks and, consequently, could increase CO_2 transportation costs and complexity.

Oil and gas production fields also present CO_2 sequestration challenges due to numerous boreholes from historical well-drilling activity. Geologists are concerned that old oil and gas wells may be inadequately sealed and that their locations may be uncertain.[19] Increased leakage risks from old wells, as well as associated mitigation and monitoring costs, may reduce the economic CCS sequestration potential in oil or gas fields. Although revenues from CO_2 sales for EOR projects could offset CO_2 transportation and sequestration costs for some source facilities, long-term CO_2 emissions in the MRCSP region would far exceed CO_2 requirements for EOR. It is possible, therefore, that because of their limited sequestration capacity and wellbore leakage concerns, oil and natural gas fields in the MRCSP region may not be viable sequestration sites for the largest CO_2 sources either.

Mt. Simon Formation

If neither the Rose Run formation nor regional coal, oil, or gas fields can provide adequate CO_2 sequestration for the major power plants in the MRCSP region, the next best potential CO_2 sequestration site is the Mt. Simon formation.

This formation is a deep saline aquifer like the Rose Run formation, but it is over four times larger in terms of sequestration capacity and is less fractured.[20]

Figure 3 shows hypothetical CO_2 pipelines which might be required if any of the major power plants in this analysis were required to transport CO_2 to the Mount Simon formation. As in the Rose Run case, **Figure 3** assumes pipelines use existing rights of way and that a 25-mile diameter, non-overlapping reserve area is needed for each plant's sequestration site. However, consistent with the Rose Run limitations, the Mt. Simon scenario assumes that the thinnest parts of the formation (the easternmost contours on the contour map) are unsuitable sequestration sites. As the figure shows, the pipelines required in such a scenario could be substantial, ranging in length from 130 to 294 miles, and averaging 234 miles. According to estimates from CMU, the approximate capital costs for these pipelines would range from $70 million to $180 million, and would average $150 million. The average levelized cost would be approximately $2.00 per ton of transported CO_2.[21]

Although **Figure 3** shows a pipeline route for all the 11 power plants in question, how many of these pipelines might be needed depends upon which plants may be able to sequester their CO_2 emissions closer to home. Furthermore, there are potential scale economies for large, integrated CO_2 pipeline networks that link many sources together rather than single, dedicated pipelines between individual sources and sequestration reservoirs.[22] The individual pipelines required in **Figure 3** may be so large on their own that combining multiple CO_2 flows from multiple plants through shared trunk lines may be limited.[23]

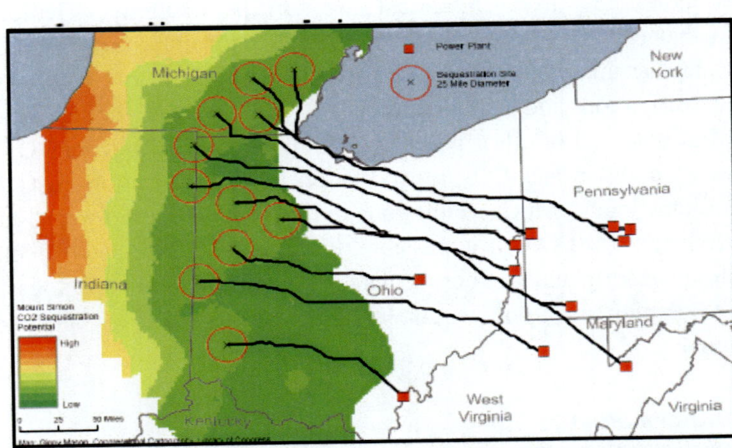

Source: MRCSP, CRS. Geologic data for western Indiana are not provided by the MRCSP.

Figure 3. Hypothetical CO_2 Pipelines to the Mt. Simon Formation

While the Mt. Simon scenario in **Figure 3** is far less favorable in terms of cost and siting requirements than the Rose Run scenario in **Figure 2**, it is not necessarily the "worst" case in terms of overall pipeline requirements. Future work on sequestration capacity may conclude that the Mt. Simon sequestration sites should be located in thicker parts of the formation (in central Indiana and Michigan) to absorb the tremendous volumes of CO_2 generated by these power plants. Such a westward shift would require even longer pipelines than those illustrated here.

POLICY IMPLICATIONS

The MRCSP pipeline scenarios, while only illustrative, nonetheless highlight several important policy considerations which may warrant congressional attention. These include concerns about CO_2 pipeline costs, siting challenges, pipeline and sequestration site relationships, and differences in sequestration potential among regions.

Variability of CO_2 Pipeline Costs

The cost of CO_2 transportation is a function of pipeline length (among other factors), which in turn is determined by the location of sequestration sites relative to CO_2 sources. The scenarios in this chapter illustrate how different assumptions about sequestration site viability in the MRCSP region can lead to a 20-fold difference in CO_2 pipeline lengths and, therefore, similarly large differences in capital costs. (In this regard, CO_2 pipeline costs may present the cost component in integrated CCS schemes with the greatest potential variability.) At the international and national policy levels, some studies have recognized this potential variability. For example, an MIT analysis states that the costs of CO_2 pipelines are highly variable due to "physical ... and political considerations."[24] The IPCC report likewise estimates total costs of CO_2 mitigation of \$31- \$71 per ton of CO_2 avoided for a new pulverized coal power plant, assuming CO_2 pipeline transportation costs, including operations and maintenance costs, of \$0 to \$5 per ton.[25] Recent increases in the global price of steel used to make line pipe could push CO_2 pipeline costs above this range.[26] At \$5 per ton of transported CO_2, pipeline costs account for a modest share of aggregate carbon control costs — between 7% and 16% based on the IPCC estimates. Nonetheless, if CCS technology were deployed on a national scale, overall CO_2 pipeline costs could be in the billions of dollars. Minimizing these costs while achieving environmental objectives may therefore be an important public policy objective.

From the perspective of individual power plants, or other CO_2 sources, highly variable costs for CO_2 pipelines may have more immediate ramifications. If CO_2 pipeline costs for specific regions reach hundreds, or even tens, of millions of dollars per plant, then power companies may have difficulty securing the capital financing or regulatory approval needed to construct or retrofit fossil fuel-powered plants in these regions. For example, in August 2007, the Minnesota Public Utilities Commission rejected a developer's proposal to construct a new coal-fired power plant in the state, in large part because the associated costs of a 450-mile CO_2 pipeline to an EOR site in Alberta, over $635 million, were not viewed to be in the public interest.[27] To the extent that other, lower-cost power plant options are available, the failure of a costly project like the Minnesota plant may not be a problem. However, if other generation sources are constrained (e.g., nuclear, renewable), then the inability to construct a new fossil-fueled power plant may negatively impact the regional balance of electricity supply and demand. Higher electricity prices or reliability concerns might ensue.

Some analysts believe that CO_2 pipeline costs will be moderated in the future because generating companies will construct new power plants geographically near sequestration sites. Recent network cost models suggest otherwise. On a mile-for-mile basis, these models show that electricity transmission costs (including capital, operations, maintenance, and electric line losses) generally outweigh CO_2 pipeline costs in new construction. Accordingly, the least costly site for a new power plant tends to be nearer the electricity consumers (cities) rather than nearer the sequestration sites if the two are geographically separated.[28] Analysts have therefore concluded that "a power system with significant amounts of CCS requires a very large CO_2 pipeline infrastructure."[29]

CO_2 Pipeline Siting Challenges

Any company seeking to construct a CO_2 pipeline must secure siting approval from the relevant regulatory authorities and must subsequently secure rights of way from landowners. There is no federal authority over CO_2 pipeline siting, so it is regulated to varying degrees by the states (as is the case for oil pipelines). The state-by-state siting approval process for CO_2 pipelines may be complex and protracted, and may face public opposition, especially inpopulated or environmentally sensitive areas.[30] Securing rights of way along existing easementsfor other infrastructure (e.g., gas pipelines), as the scenarios in this chapter assume, may be one way to facilitate the siting of new CO_2 pipelines. However, questions arise as to the right of easement holders to install CO_2 pipelines, compensation for use of such easements, and whether existing easements can be sold or leased to CO_2 pipeline companies.[31] Although these siting issues may arise for any CO_2 pipeline, they become more challenging as pipeline systems become larger and more interconnected, and cross state lines. If a widespread, interstate CO_2 pipeline network is required to support CCS, the ability to site these pipelines may become an issue requiring new federal initiatives.[32]

Pipeline and Sequestration Site Relationships

Due to potential CO_2 transportation costs, individual generating plants have a strong interest in the selection of specific sequestration sites under future CCS policies. Since transporting CO_2 to distant locations can impose significant additional costs to a facility's carbon control infrastructure, facility owners may seek regulatory approval for as many sequestration sites as possible and near to as many facilities as possible. Furthermore, capacity limitations at favorably located sequestration sites (like the Rose Run formation) may lead to competition among large CO_2 source facilities seeking to secure the best local sequestration sites before others do. How the development of sequestration sites will be prioritized and how competition for such sites may evolve have yet to be explored, but they may create new and significant economic differences among facilities.

Because CO_2 pipeline requirements in a CCS scheme are driven by the relative locations of CO_2 sources and sequestration sites, identification and validation of such sites must explicitly account for CO_2 pipeline costs if the economics of those sites are to be fully understood. Proposals such as S. 2323, which would require an integrated evaluation of CO_2 capture, sequestration, and transportation (Sec. 3(b)(5)), appear to promote such an approach, although the details of future sequestration site selection have yet to be established. If CCS moves from pilot projects to widespread implementation, government agencies and private companies may face challenges in identifying, permitting, developing, and monitoring the large number of localized sequestration reservoirs that may be proposed.

Advantaged and Disadvantaged Regions

Geologists have long recognized that some regions in the United States have high potential for carbon sequestration and others do not. For example, a 2007 study at Duke University concluded that "geologic sequestration is not economically or technically feasible within North Carolina," but "may be viable if the captured CO_2 is piped out of North Carolina and stored elsewhere."[33] Likewise, states in the Northeast, Minnesota, Wisconsin, and possibly parts of other states appear to lack geological formations with potential for large-scale sequestration of the volumes of CO_2 they produce. If national CCS policies are implemented, power plants and other CO_2-producing facilities in these states may face more extensive, and more costly, pipeline requirements than other states if they are to sequester their CO_2. States such as North Carolina, with limited sequestration potential and a relatively high proportion of coal or natural gas in their electric generation fuel mix, may face particular challenges in this regard. The Duke study, for example, estimated it would cost $5 billion to construct an interstate pipeline network for transporting CO_2 from North Carolina's electric utilities to sequestration sites in other states.[34]

One particular concern among some stakeholders is that high CO_2 transportation costs could increase electricity prices in "sequestration-poor" regions relative to regions able to sequester CO_2 more locally. For states like Massachusetts, for example, which has some of the highest electricity prices in the country and may have little sequestration potential, CO_2 transportation costs could raise electricity prices even higher above the national average. Moving beyond this illustrative example to evaluate comprehensively the distribution of CO_2 transportation costs across the United States is beyond the scope of this chapter. Nonetheless, these kinds of regional price impacts, and their implications for regional economies, may become an issue for Congress.

CONCLUSION

The socially and economically efficient development of the nation's public infrastructure is an important consideration for policymakers. In the context of a national program for CCS, CO_2 pipelines may be a major addition to this infrastructure. Yet there are many uncertainties about the cost and configuration of CO_2pipelines thatwould be needed to meet environmental goals withinan emerging regulatory framework. Exactly who will pay for CO_2 pipelines, and how, is beyond the scope of this chapter, but understanding ways to minimize the cost and environmental impact of this infrastructure may be of benefit to all.

In addition to specific questions about CO_2 pipeline requirements, the scenarios in this chapter raise larger questions about the ultimate development and allocation of sequestration capacity under a national CCS policy. How much individual companies may have to spend to transport their CO_2 depends upon where it has to go. However, even as viable sequestration reservoirs are being identified, it is unclear which CO_2 source facilities will have access to them, under what time frame, and under what conditions. While Congress is beginning to turn its attention to these questions, it will likelyrequire sustained attention and the input of many stakeholders to refine and address them. Given the potential size of a national CO_2 pipeline network, many billions of dollars of capital investment may be affected by policy decisions made today.

End Notes

[1] See, for example: John Deutch, Ernest J. Moniz, et al., *The Future of Coal.* (Cambridge, MA: Massachusetts Institute of Technology: 2007): 58. (Hereafter referred to as MIT 2007.)

[2] Intergovernmental Panel on Climate Change, Special Report: *Carbon Dioxide Capture and Storage, 2005* (2005): 190. (Hereafter referred to as IPCC 2005.)

[3] Eric Williams, Nora Greenglass, and Rebecca Ryals, "Carbon Capture, Pipeline and Storage: A Viable Option for North Carolina Utilities?" Working paper prepared by the Nicholas Institute for Environmental Policy Solutions and the Center on Global Change, Duke University (Durham, NC: March 8, 2007): 4.

[4] See, for example: BP, "Right on the Route," *Frontiers*, Issue 19 (August, 2007). [http://www.bp.com/sectiongenericarticle.do?categoryId=9019307&contentId=7035194]

[5] R.T. Dahowski, J.J. Dooley, C.L. Davidson, S. Bachu, N. Gupta, and J. Gale, "A North American CO_2 Storage Supply Curve: Key Findings and Implications for the Cost of CCS Deployment," *Proceedings of the Fourth Annual Conference on Carbon Capture and Sequestration* (Alexandria, VA: May 2-5, 2005). The study addresses CO_2 capture at 2,082 North American facilities including power plants, natural gas processing plants, refineries, cement kilns, and other industrial plants.

[6] Jennie C. Stevens and Bob Van Der Zwaan, "The Case for Carbon Capture and Storage," *Issues in Science and Technology*, vol. XXII, no. 1 (Fall 2005): 69-76. (See page 15 of this chapter for a discussion of safety issues.)

[7] U.S. Dept. of Energy, Office of Fossil Energy, *Carbon Sequestration Atlas of the United States and Canada*, (2007).

[8] National Energy Technology Laboratory, "NatCarb" online database, March 29, 2007. [http://drysdale.kgs.ku.edu/natcarb/med_view/map.cfm?kid=242&theme_kid=3]

[9] For most of the area shown in **Figure 1**, the Rose Run sandstone lies at depths greater than 2,500 feet — deep enough to make the formation potentially suitable for CO_2 sequestration.

[10] Model found in: Sean T. McCoy and EdwardS. Rubin, "An Engineering-Economic Model of Pipeline Transport of CO_2 with Application to Carbon Capture and Storage," *International Journal of Greenhouse Gas Control*, In press (November 19, 2007). Costestimates were provided by Sean McCoy at the request of CRS.

[11] M.D. Zoback, H. Ross, and A. Lucier, "Geomechanics and CO_2 Sequestration," *GCEP Technical Report 2006*, Stanford Univ., Global Climate and Energy Project (2006):11. [http://gcep.stanford.edu/pdfs/QeJ5maLQQrugiSYMF3ATDA/2.4.2.zoback_06.pdf]

[12] U.S. Dept of Energy, Office of Fossil Energy, *Carbon Sequestration Atlas of the United States and Canada*, (2007):38.

[13] Amie Lucier, Mark Zoback, Neeraj Gupta, and T. S. Ramakrishnan, "Geomechanical Aspects of CO_2 Sequestration in a Deep Saline Reservoir in the Ohio River Valley Region," *Environmental Geosciences* (June 2006), 13(2):85-103.

[14] S. Julio Friedmann, *Site Characterization and Selection Guidelines for Geological Carbon Sequestration*, Lawrence Livermore National Laboratory, UCRL-TR-234408 (September 7, 2007); K. Prasad Saripalli, B. Peter McGrail, and Mark D. White, "Modeling the Sequestration of CO2 in Deep Geological Formations," *Proceedings of the First National Conference on Carbon Sequestration*, National Energy Technology Laboratory (May 14-17, 2001):11.

[15] For an example of such research, see: L. Chiaramonte, M. Zoback, M., S.J. Friedmann, and V. Stamp, "Seal Integrity and Feasibility of CO_2 Sequestration in the Teapot DomeEOR Pilot: Geomechanical Site Characterization," *Environmental Geoscience*, v. 53 (2007).

[16] According to the DOE Carbon Sequestration Atlas (pp. 38-39), there are one billion metric tons of total potential capacity for CO_2 in coal seams versus nearly 20 billion metric tons for the Rose Run sandstone.

[17] Cui, X., R. M. Bustin, and L. Chikatamarla, "Adsorption-induced Coal Swelling and Stress: Implications for Methane Production and Acid Gas Sequestration into Coal Seams," *Journal of Geophysical Research*, vol. 112, B10202 (2007).

[18] U.S. Dept of Energy (2007):36.

[19] IPCC 2005:215; Charles W. Zuppann, "Too Much Fun? — Tales of 'Field Checking' at the Indiana Geological Survey," *The PGI Geology Standard,* No. 48 (April 2007): 6-9. [http://www.indiana.edu/~pgi/docs/Standard%20Issues/PGI_Issue_48.pdf]

[20] U.S. Dept of Energy, (2007):38.

[21] Sean T. McCoy and Edward S. Rubin (November 19, 2007). Cost estimates were provided by Sean McCoy at the request of CRS.

[22] MIT 2007: 58.

[23] Gemma Heddle, Howard Herzog, and Michael Klett, "The Economics of CO_2 Storage," MIT Laboratory for Energy and the Environment, Working Paper MIT LFEE 2003-003 RP (August 2003): 23.

[24] MIT 2007: 58.

[25] IPCC report: 347.

[26] For further information about steel prices, see CRS Report RL32333, *Steel: Price and Policy Issues*, by Stephen Cooney.

[27] Minnesota Public Utilities Commission, *Order Resolving Procedural Issues, Disapproving Power Purchase Agreement, Requiring Further Negotiations, and Resolving to Explore the Potential for a Statewide Market for Project Power under Minn. Stat. § 216b.1694, Subd. 5*, Docket No. E-6472/M-05-1993 (August 30, 2007):15; Minnesota Public Utilities Commission, *Staff Briefing Papers — Appendix I*, Docket No. E-6472/M-05-1993 (July 31, 2007):78. Cost estimate in 2011 U.S. dollars.

[28] Jeffrey M. Bielicki and Daniel P. Schrag, "On the Influence of Carbon Capture and Storage on the Location of Electric Power Generation," Harvard University, Belfer Center for Science and International Affairs, Working paper (2006).

[29] Adam Newcomer and Jay Apt, "Implications of Generator Siting for CO_2 Pipeline Infrastructure," Carnegie Mellon Electricity Industry Center, Working Paper CEIC-07-11 (2007).

[30] National Commission on Energy Policy, *Siting Critical Energy Infrastructure: An Overview of Needs and Challenges*. (Washington, DC: June 2006): 9. (Hereafter referred to as NCEP 2006.)

[31] Partha S. Chaudhuri, Michael Murphy, and Robert E. Burns, "Commissioner Primer: Carbon Dioxide Capture and Storage" (National Regulatory Research Institute, Ohio State Univ., Columbus, OH: March 2006): 17.

[32] For further discussion, see CRS Report RL34307, *Regulation of Carbon Dioxide (CO_2) Sequestration Pipelines: Jurisdictional Issues*, by Adam Vann and Paul W. Parfomak.

[33] Williams, et al., (2007): 4.

[34] Williams, et al., (2007): 20.

In: Pipelines for Carbon Sequestration: Background ... ISBN: 978-1-60741-383-7
Editors: Elvira S. Hoffmann

Chapter 3

REGULATION OF CARBON DIOXIDE (CO_2) SEQUESTRATION PIPELINES: JURISDICTIONAL ISSUES

Adam Vann[1] and Paul W. Parfomak[2]
[1]Legislative Attorney, American Law Division
[2] Specialist in Energy and Infrastructure Policy Resources, Science, and Industry Division

SUMMARY

In the last few years there has been a surge in interest in carbon capture and sequestration (CCS) as a way to mitigate manmade carbon dioxide (CO_2) emissions and thereby help address climate change concerns. This approach may require the transportation of captured carbon dioxide via pipeline to an underground storage reservoir with the intent of keeping it out of the atmosphere. Congress has begun to address this issue.

The recently enacted Energy Independence and Security Act of 2007 (P.L. 110-140) contains measures to promote research and development of CCS technology, to assess sequestration capacity, and to clarify the framework for issuance of CO_2 pipeline rights-of-way on public land. Other legislative measures, including S. 2191 and S. 2323, have also sought to encourage the development of CO_2 sequestration, capture, and transportation technology. Further, measures have been introduced in Congress, including the Carbon Dioxide Pipeline Study Act of

2007 (S. 2144), that would require the Secretary of Energy to study the feasibility of constructing and operating a network of CO_2 pipelines for CCS. If these pipelines crossed state lines, it could raise important issues concerning regulatory jurisdiction over siting and pricing.

This chapter discusses federal jurisdictional uncertainty over CO_2 pipelines under existing law, as well as potential legislative activity to address any possible regulatory "gap" in jurisdiction. It should be noted that such a potential gap does not necessarily demand federal legislative resolution. Nevertheless, some analysts, drawing on the history of oil and natural gas pipeline development, anticipate a potential need for better defined federal legislative/regulatory authority over an interstate CO_2 pipeline network. Accordingly, Congress may be called upon to consider whether existing federal jurisdictional disclaimers and the state-by-state regulatory structure for Enhanced Oil Recovery (EOR) pipelines is an appropriate regulatory scheme for a possible interstate CO_2 pipeline network in support of CCS. Congress could opt to amend existing statutes, possibly those related to the interstate pipeline jurisdiction of the Federal Energy Regulatory Commission (FERC) or the Surface Transportation Board (STB) to provide for definitive CO_2 pipeline rate jurisdiction by the federal government if it does not favor the existing regulatory scheme. Alternatively, Congress could establish another federal regulator of CO_2 pipelines or possibly other legislation to amend the existing regulatory scheme.

Introduction

In the last few years there has been a surge in interest in carbon capture and sequestration (CCS) as a way to reduce carbon dioxide (CO_2) emissions, and thereby help to address concerns about climate change. The recently enacted Energy Independence and Security Act of 2007, P.L. 110-140, contains measures to promote research and development of CCS technology and assess sequestration capacity, including a measure to clarify the framework for issuance of CO_2 pipeline rights-of-way on public land. Other legislative proposals, including S. 2191 and S. 2323, have also sought to encourage the development of CO_2 sequestration, capture, and transportation technology. Further, measures have been introduced in Congress, including the Carbon Dioxide Pipeline Study Act of 2007 (S. 2144), which would require the Secretary of Energy to study the feasibility of constructing and operating a network of CO_2 pipelines for CCS.

If interstate CO_2 pipelines for carbon sequestration are ultimately developed, they could raise a variety of issues concerning jurisdiction over the siting of these pipelines and the prices they may charge for their transportation services. While jurisdiction over CO_2 pipeline safety resides with the Office of Pipeline Safety (a branch of the Department of Transportation), jurisdiction over hypothetical interstate CO_2 pipeline siting and rate decisions is not clear. Jurisdiction could fall to the Federal Energy Regulatory Commission (FERC) or to the Surface Transportation Board (STB). However, both agencies have at some point expressed a position that interstate CO_2 pipelines are not within their purview.

This chapter reviews the history of pipeline regulation, including the limited history of interstate CO_2 pipeline regulation, and examines the regulatory missions of FERC, the STB, and other agencies. The report discusses possible responses to preceived jurisdictional uncertainties under existing law as well as potential legislative steps intended to address any potential regulatory "gap" in interstate CO_2 pipeline jurisdiction.

HISTORY AND BACKGROUND

The transportation of CO_2 via pipeline is not a hypothetical concept, nor even a recent development. For many years companies have moved CO_2 from one location to another for enhanced oil recovery (EOR) — a process which extends oil production from mature oil fields. There are several EOR methods, but gas injection, dominated by CO_2 injection, accounts for nearly 50% of EOR production in the United States.[1] During this process, CO_2 is injected underground at an oil field, forcing more oil from the reservoir to the extraction wells, ultimately increasing the amount of oil that can be recovered from a given field.

Approximately 5,800 kilometers (3,600 miles) of CO_2 pipeline operate today in the United States.[2] The oldest long-distance CO_2 pipeline in the United States constructed for EOR is the 225 kilometer Canyon Reef Carriers Pipeline (in Texas), which began service in 1 972.[3] Other large CO_2 pipelines constructed since then, mostly in the Western United States, have expanded the CO_2 pipeline network for EOR. These pipelines carry CO_2 from naturally occurring underground reservoirs, natural gas processing facilities, ammonia manufacturing plants, and a large coal gasification project to regional oil fields. Additional pipelines may carry CO_2 from other manmade sources to supply a range of industrial applications. However, with one notable exception discussed below, federal regulation of siting and rates for these pipelines has not been addressed, due in large part to the fact that many of them are intrastate and that they often transport CO_2 for the benefit of the pipeline's owners (so there are no rate or service disputes).

Recent Carbon Control Activity

In recent years, the public, the press, and government officials have shown increasing concern about global climate change. Many have suggested that recent increases in the global temperature, as well as future anticipated increases, are the result, at least in part, of emissions of large quantities of CO_2 from manmade sources. One of the more prominent ideas to limit CO_2 emissions, especially CO_2 emissions from power plants, is carbon capture and direct sequestration (CCS). CCS is a process whereby CO_2 emissions are "captured" at their source and then stored ("sequestered") either underground or elsewhere, rather than being released into the atmosphere.[4]

CCS science and associated technology are still in the early stages of development. Among many questions yet to be answered is whether we may ultimately see a large number of sequestration sites located geographically close to CO2 source facilities, or a smaller number of more centralized, or more distant, sequestration locations.[5] If a widespread CCS network does develop, whether that network features decentralized or centralized storage configurations remains to be seen; however, pipeline requirements and storage configurations are necessarily related. If sequestration is widely used and relatively centralized, a significant interstate pipeline network would likely be required to transport the CO_2 to the sequestration sites.

FERC and STB Pipeline Jurisdiction

If an interstate pipeline system for CCS more extensive and elaborate than the current pipeline network for EOR is to be developed, questions arise as to who will regulate pipeline siting and the rates to be charged for the transportation of CO_2. Based on their current regulatory roles, two of the more likely candidates for jurisdiction over interstate pipelines transporting CO_2 for purposes of CCS are the Federal Energy Regulatory Commission (FERC) and the Surface Transportation Board (STB).

The Natural Gas Act of 1938 (NGA) vests in FERC the authority to issue "certificates of public convenience and necessity" for the construction and operation of interstate natural gas pipeline facilities.[6] FERC is also charged with extensive regulatory authority over the siting of natural gas import and export facilities, as well as rates for transportation of natural gas and other elements of transportation service. FERC also has jurisdiction over regulation of oil pipelines pursuant to the Interstate Commerce Act (ICA).[7] The ICA, as amended by the Hepburn Act of 1905, provided that the Interstate Commerce Commission (ICC) was to have jurisdiction over rates and certain other activities of interstate oil pipelines, as these pipelines were considered to be "common carriers."[8] This jurisdiction was transferred to FERC in the Department of Energy Organization Act of 1977.[9] FERC's jurisdiction over oil pipelines is not as extensive as its jurisdiction over natural gas pipelines. FERC is not involved in the oil pipeline siting process. However, as with natural gas, FERC does regulate transportation rates and capacity allocation for oil pipelines.[10]

FERC is assigned jurisdiction over natural gas and oil pipelines explicitly in statutory language. Jurisdiction over rate regulation for "other" types of pipelines resides with the STB. The STB is an independent regulatory agency (administratively affiliated with the Department of Transportation) charged by Congress with the primary mission of resolving railroad disputes pursuant to the ICA. It is the successor agency to the ICC. Pipelines, like railroads, are "common carriers" used by more than one company for the transportation of goods. Therefore, the ICA also assigned the ICC (and thus the STB) oversight authority over pipelines transporting a commodity other than "water, gas or oil."[11] It should be noted, however, that unlike FERC, the STB does not require pipeline companies to file tariffs and justify their rates. Instead, the STB acts as a forum for resolution of disputes related to pipelines within its jurisdiction. Parties who wish to challenge a rate or another aspect of a pipeline's common carrier service may petition the STB for a hearing; there is no ongoing regulatory oversight.

Thus, there are two federal regulatory agencies that, generally speaking, have jurisdiction over interstate pipeline rate and capacity allocation matters. However, as explained below, both of these agencies appear to have explicitly rejected jurisdiction over CO_2 siting and rates, and there is no legislative or judicial history indicating that their rejections were improper. These decisions, the reasoning behind them, and the status of federal jurisdiction over CO_2 pipelines are covered in the next section.

Prior CO_2 Pipeline Regulation: Cortez Pipeline

As mentioned above, CO_2 is used for enhanced oil recovery, and in some cases companies have used pipelines to transport CO_2 over state lines (i.e., in interstate commerce) for EOR purposes. One such pipeline of regulatory significance is the Cortez Pipeline, which runs through Colorado, New Mexico, and Texas.

FERC Decision

In December of 1978, the Cortez Pipeline Company (Cortez) sought a declaratory order from FERC that the construction and operation of a proposed interstate pipeline transporting a gas comprising of 98% CO_2 and 2% methane would not be within the Commission's jurisdiction. Cortez argued that the gas in question was not "natural gas" as the term is defined in Section 2(5) of the NGA,[12] so a proposed pipeline to transport this gas was not under FERC's NGA jurisdiction. FERC agreed with Cortez and issued a declaratory order disclaiming jurisdiction over the proposed pipeline.[13] In its decision, FERC explored the inherent ambiguity in the term "natural gas," explaining that it has two very different definitions. FERC recognized that in the terminology of chemistry, "natural gas" refers to any substance that is gaseous in its natural state, including carbon dioxide.[14] However, according to FERC, the more common usage of the term "natural gas" refers to a gaseous mixture of hydrocarbons.[15] FERC held that it was this meaning of "natural gas" that applied to the term as it was used in the NGA. FERC pointed to the goals and purposes of the NGA, which are primarily to regulate a specific "natural gas" industry.[16] Thus, the term "natural gas" as used in the statute referred to a gaseous mixture of hydrocarbons.[17] As a result, FERC held that the proposed Cortez Pipeline was not within the NGA jurisdiction of the Commission.[18]

ICC Decision

In 1980, after FERC issued its CO_2 ruling, the owners of the proposed Cortez Pipeline went to the ICC to seek a similar declaratory order that the pipeline would not be subject to the ICC's jurisdiction either. As the ICC recognized at the outset of its decision, there was no controversy concerning whether ICC approval was necessary for construction or expansion of pipeline facilities — the statute and previous case law plainly state that the ICC has no pipeline siting jurisdiction whatsoever.[19] Furthermore, the ICC noted that the U.S. Department of Transportation would exercise jurisdiction over the pipeline's compliance with applicable safety standards.[20] Instead, the decision focused solely upon the ICC's regulation of other aspects of pipeline service (i.e., rates), and whether CO_2 pipelines are covered by the statutory exception from ICC regulation for pipelines that transport "water, gas or oil."[21]

Relying on the legislative history of previous versions of the statutory language in the ICA (which excluded "natural or artificial gas") as well as the plain language of the current statute, the ICC concluded that Congress intended to exclude all types of gas, including CO_2, from ICC regulation. The ICC recognized that its initial ruling in this matter, in concert with FERC's order disavowing jurisdiction over the proposed Cortez Pipeline, created a regulatory gap of sorts. The ICC noted that generally, "[t]he opinion of a sister agency should be given weight, if possible, so that related statutes can be coordinated."[22] However, the ICC found that "in this case the FERC decision is not helpful to us because it did not construe or interpret the terms natural and artificial gas [under the ICA]. Its decision was based on other grounds."[23] Although the ICC found in this initial decision that it likely did not have jurisdiction over CO_2 pipelines, it did conclude that "the issue is important enough to institute a proceeding and accept comments on the petition and our view on it."[24]

After the comment period the ICC confirmed its view that CO_2 pipelines were excluded from the ICC's jurisdiction.[25]

Potential Issues Related to ICC Jurisdiction

Notwithstanding the ICC's 1980 disclaimer of jurisdiction over CO_2 pipelines in the Cortez Pipeline case, other evidence indirectly suggests the possibility that interstate CO_2 pipelines could still be considered subject to the jurisdiction of the STB. For example, an April 1998 report by the General Accounting Office (GAO)[26] stated that interstate CO_2 pipelines, as well as pipelines transporting other gases, are subject to the board's oversight authority. The report stated that GAO had identified five products carried by 21 pipelines subject to the STB ' s jurisdiction.[27] One of the five identified products was CO_2. (Another was hydrogen — also a gas). In fact, the report lists 14 different pipelines transporting CO_2 for purposes of EOR, including the Cortez Pipeline, which are said to be subject to the jurisdiction of the STB.[28] The GAO states that the STB reviewed its analysis and, presumably, did not object to this jurisdictional classification.[29]

It should also be noted that the Cortez Pipeline decision was issued by the ICC, not the STB. Although the STB is the successor to the now-defunct ICC, and the statutory language regarding the STB's jurisdiction is virtually identical to the language at issue in the Cortez decision, they are not the same agency. The STB conceivably could determine that its jurisdiction is not governed by the ICC's decision in the Cortez Pipeline matter. Indeed, the Supreme Court has ruled that federal agencies are not precluded from changing their positions on the issue of regulatory jurisdiction. According to the Court, "an initial agency interpretation is not instantly carved in stone. On the contrary, the agency, to engage in informed rulemaking, must consider varying interpretations and the wisdom of its policy on a continuing basis."[30] Accordingly, regulation of CO_2 pipelines for CCS purposes by the STB (or by FERC, for that matter) under existing statutes remains a possibility.

POLICY IMPLICATIONS OF THE POSSIBLE REGULATORY "GAP"

If CCS technology develops to the point where interstate CO_2 pipelines become more common, and if FERC and the STB continue to disclaim jurisdiction over CO_2 pipelines, then the potential regulatory "gap" discussed above may receive attention. This gap does not necessarily demand resolution. As commenters have noted, state laws and contractual arrangements among interested parties established under the EOR model would also apply to CO_2 pipelines for CCS.[31] Interstate CO_2 pipelines would still be required to meet the safety requirements of the Department of Transportation. Also, although there would be no federal rate regulation, any anticompetitive behavior by the owners or operators of a CO_2 pipeline could be addressed by federal antitrust enforcement agencies, including the Federal Trade Commission and the antitrust division of the U.S. Department of Justice. Finally, any pipelines needing to cross federal lands would be required to obtain a right-of-way from the federal government, and so would be subject to any conditions such rights-of-way might impose.

Nevertheless, some analysts, drawing on the history of oil and natural gas pipeline development, anticipate a potential need for better defined federal regulatory authority over a potential expansive CO_2 pipeline network.[32]

> The growth of these sectors and the growing importance of transportation, abuses of market power, price discrimination and other issues, including conflicts between state and federal authorities led both to preemption by the federal government and administrative regulation. As the CO_2 transport and storage sector grows, similar issues of regulatory frameworks and the mix of federal and state jurisdiction are likely to have to be confronted, as has been the case for all network industries in the United States. The eventual economic regulatory development for CCS will need to consider the varying approaches taken for oil and natural gas, and the serious problems that their history experienced.

Thus, Congress may ultimately be asked to consider whether the existing federal jurisdictional disclaimers and the state-by-state regulatory structure for EOR pipelines is an appropriate regulatory scheme for a potential national CO_2 pipeline network in support of CCS. Congress could choose to amend existing statutes to provide for definitive CO_2 pipeline rate jurisdiction by the federal government if it does not favor the existing regulatory scheme. FERC and the STB are two candidates for administration of such oversight. Alternatively, Congress could establish another federal regulator of CO_2 pipelines, or enact legislation addressing specific aspects of the existing regulatory structure for EOR.

End Notes

[1] U.S. Dept. of Energy, "Enhanced Oil Recovery/CO_2 Injection," web page (June 12, 2007). [http://www.fossil

[2] U.S. Dept. of Transportation, National Pipeline Mapping System database (June 2005). [https://www.npms.phmsa.dot.gov].

[3] Kinder Morgan CO_2 Company, "Canyon Reef Carriers Pipeline (CRC)," web page (2007). [http://www.kindermorgan.com/business/co2/transport _canyon_reef.cfm].

[4] For a detailed discussion of CCS, *see* CRS Report RL33801, *Carbon Capture and Sequestration (CCS)*, by Peter Folger.

[5] See, alternative views in: R.T. Dahowski, J.J. Dooley, C.L. Davidson, S. Bachu, N. Gupta, and J. Gale, "A North American CO2 Storage Supply Curve: Key Findings and Implications for the Cost of CCS Deployment," *Proceedings of the Fourth Annual Conference on Carbon Capture and Sequestration* (Alexandria, VA: May 2-5, 2005); and Jennie C. Stevens and Bob Van Der Zwaan, "The Case for Carbon Capture and Storage," *Issues in Science and Technology*, vol. XXII, no. 1 (Fall 2005): 69-76.

[6] 15 U.S.C. 717f(c).

[7] 49 App. U.S.C.§1.

[8] Id. at 1(1), 1(4), and 1(7).

[9] P.L. 95-224.

[10] Section 1801 of the Energy Policy Act of 1992 directed FERC to "promulgate regulations establishing a simplified and generally applicable ratemaking methodology" for oil pipeline transportation.

[11] 49 U.S.C. § 1-501(a)(1)(c).

[12] 15 U.S.C. § 717(a)(5).

[13] Cortez Pipeline Company, 7 FERC ¶ 61,024 (1979).

[14] Id. at 61,041.

[15] Id.

[16] Id.

[17] Id.

[18] Id. at 61,042.

[19] Cortez Pipeline Company — Petition for Declaratory Order — Commission Jurisdiction Over Transportation of Carbon Dioxide by Pipeline, 45 Fed. Reg. 85177 (December 24, 1980).

[20] Id.

[21] 49 U.S.C. § 1-501(a)(1)(c).

[22] Cortez Pipeline Co., 45 Fed. Reg. at 85178.

[23] Id. at 85178.

[24] Id.

[25] Cortez Pipeline Company — Petition for Declaratory Order — Commission Jurisdiction Over Transportation of Carbon Dioxide by Pipeline, 46 Fed. Reg. 18805 (March 26, 1981).

[26] Now known as the Government Accountability Office.

[27] GAO Report: SURFACE TRANSPORTATION: Issues Associated With Pipeline Regulation by the Surface Transportation Board, April 1998.

[28] Id. at 27.

[29] Surface Transportation Board (STB), Personal communication, (December 2007). The STB Office of Governmental and Public Affairs informed CRS that the board recognizes the conflict between this GAO report and the ICC decision (as well as the wording of 49 C.F.R. § 15301 governing STB jurisdiction over pipelines other than those transporting "water, gas or oil"). However the office did not want to state an opinion as to the current extent of STB jurisdiction over CO_2 pipelines and suggested that the STB would likely not act to resolve this conflict unless a CO_2 pipeline dispute comes before it.

[30] Chevron U.S.A. v. Nat. Res. Def. Council, 467 U.S. 837, at 863-64 (1984).

[31] See, for example, Philip M. Marston, "Doing the Deal: Legal and Regulatory Aspects of the Evolving CCS Regime in the USA," Proceedings of the *2e Collogue International Captage et Stockage Géologique du CO2*, Paris, France (October 4-5, 2007), at: 3. [http://www.colloqueco2.com/presentations2007/ColloqueCO2-2007_Session4_1 - MARSTON.pdf].

[32] M.A. de Figueiredo, H.J. Herzog, P.L. Joskow, K.A. Oye, and D.M. Reiner, "Regulating Carbon Dioxide Capture and Storage," MIT Center for Energy and Environmental Policy Research, Working Paper 07-003 (April 2007), at: 5.

In: Pipelines for Carbon Sequestration: Background … ISBN: 978-1-60741-383-7
Editors: Elvira S. Hoffmann

Chapter 4

REGULATORY ASPECTS OF CARBON CAPTURE, TRANSPORTATION AND SEQUESTRATION

U.S. Government Printing Office

OPENING STATEMENT OF HON. JEFF BINGAMAN, U.S. SENATOR FROM NEW MEXICO

The Chairman. All right. Why don't we get started here? I'm informed Senator Domenici is on his way.

We have two of our colleagues here to talk first about the legislation that is the subject of our hearing. Let me give a very short opening statementand then turn to them.

I'd like to welcome everybody and thank the witnesses who are testifying before the committee. This is a legislative hearing on two bills, S. 2144, that Senator Coleman and some others have introduced, and S. 2323, that Senator Kerry has introduced along with several of our colleagues.

These two bills focus on important policy aspects of carbon dioxide capture, transportation and storage. S. 2144 focuses on the issue of expanding the existing carbon dioxide pipeline infrastructure. S. 2323 focuses more broadly on carbon capture and storage research, development and demonstration projects and perhaps more pertinent to today's hearing also focuses on developing a policy framework for rapid implementation of integrated carbon dioxide capture and storage systems.

The topic of reducing greenhouse gases, particularly carbon dioxide emissions is a topic of great concern to myself and to all members of this committee. Carbon capture and geologic storage holds promise as a measure that can be used to mitigate global climate change while still allowing the use of fossil fuels at electricity generating plants and industrial facilities. Discussion centered on coal use in a carbon constrained world, integrated carbon capture and storage systems may present the most immediate solution for continued use of coal than other carbon intensive fuels while not contributing further to carbon dioxide emissions and global warming.

Last December a historic piece of legislation was passed into law, the Energy Independence and Security Act of 2007. It included key provisions for expanding critical research and development programs aimed at bringing integrated carbon capture and storage systems to the full technological deployment stage. The new law is important for focusing research and development efforts on technologies that are essential for reducing carbon dioxide emissions.

This legislation was a first step, a key first step, in advancing carbon capture and storage projects, but additional legislation will be needed to advance these storage projects into full commercial deployment. The next phase in fast tracking deployment of these technologies is establishing a policy framework that will assist early industry movers in selecting the appropriate geologic storage sites, in operation of their facilities and in managing the facilities for decades following the closure of a geologic storage operation. The aim of this hearing is to receive testimony on these two bills and their contribution to developing a carbon dioxide capture, transport and storage policy framework.

Let me defer to Senator Domenici for any comments he has before I call on Senator Coleman and Senator Kerry for their comments.

Senator Domenici. Mr. Chairman, considering the time I would ask that you let the two witnesses, the two Senators give their remarks and then I will give mine.

The Chairman. Alright. We will proceed that way. There's no particular order here. Senator Coleman, you were the first one here and Senator Kerry is the taller of the two. Which of you would like to go?

[Laughter.]

Senator Coleman. I'll certainly defer to my senior colleague, Senator Kerry.

The Chairman. Senator Kerry, go right ahead, please. Thank you for being here.

[The prepared statements of Senators Bingaman and Salazar follow:]

Prepared Statement of Hon. Jeff Bingaman, U.S. Senator from New Mexico

I'd like to welcome everyone here today and thank the witnesses who are testifying before the committee for this legislative hearing on bills S. 2144 and S. 2323. These two bills focus on important policy aspects of carbon dioxide capture, transportation, and storage. S. 2144 focuses on the issue of expanding the existing carbon dioxide pipeline infrastructure. S. 2323 focuses more broadly on carbon capture and storage research, development and demonstration projects, and perhaps more pertinent to today's hearing it also focuses on developing a policy framework for rapid implementation of integrated carbon dioxide capture and storage systems.

The topic of reducing greenhouse gases, particularly carbon dioxide emissions, is a topic of great concern to myself and the members of this committee. Carbon capture and geologic storage holds promise as a measure that can be used to mitigate global climate change, while still allowing the use of fossil fuels at electricitygenerating plants and industrial facilities. With discussion centered on coal use in a carbon-constrained world, integrated carbon capture and storage systems may present the most immediate solution for continued use of coal and other carbon intensive fuels while not contributing further to carbon dioxide emissions and global warming.

Last December a historic piece of legislation was passed into law, the Energy Independence and Security Act of 2007, which included key provisions for expanding critical research and development programs, aimed at bringing integrated carbon capture and storage systems to the full technological deployment stage. The new law is important for focusing research and development efforts on technologies that are essential for reducing carbon dioxide emissions. This legislation was a key first step in advancing carbon capture and storage projects, but additional legislation will be needed to advance these storage projects into full commercial deployment.

The next phase in fast-tracking deployment of these technologies is establishing a policy framework that will assist early industry movers in selecting the appropriate geologic storage sites, operation of their facilities, and managing the facilities for decades following the closure of a geologic storage operation. The aim of this hearing is to receive testimony on S. 2144 and S. 2323 and their contribution to developing a carbon dioxide capture, transport, and storage policy framework.

I would like to begin the hearing by welcoming the original bill sponsors who have come to speak on the bills today, Senator Kerry will speak on S. 2323 and Senator Coleman will speak on S. 2144.

Prepared Statement of Hon. Ken Salazar, U.S. Senator from Colorado

Thank you Chairman Bingaman and Ranking Member Domenici for holding this hearing on the regulatory aspects of carbon capture, transportation, and sequestration.

Capturing carbon dioxide at its source and safely storing it to avoid its release into the atmosphere will be essential to reducing greenhouse gas emissions. I believe carbon capture and storage (CCS) should be a top priority in our nation's energy policy. There is little doubt that a successful domestic CCS program will boost our nation's coal industry, and that a low-carbon footprint coal industry is critical to our nation's energy and environmental security.

To make CCS an effective, reliable, and cost-feasible reality, we need to move forward simultaneously on two fronts: we need to aggressively develop both the technical knowledge necessary and the regulatory framework for CCS infrastructure development.

On the technical front, I sponsored the provision of the new energy bill that directs the United States Geological Survey and the Departments of Energy and the Interior to coordinate a national assessment of our carbon dioxide storage capacity. I also fought to include the provisions that will expand DOE's CCS research and development programs, with a particular eye towards the large-scale CCS demonstration projects that are crucial to achieving commercial viability. I am looking forward to learning about DOE's most recent progress today.

I am glad that today's hearing will focus attention on the second front—the regulatory front. We need to establish a regulatory framework for the transport and storage of carbon dioxide. As you know I am an original cosponsor of S. 2144, the Carbon Dioxide Pipeline Study Act of 2007, which would instruct the federal agencies present today to perform a broad feasibility study of the construction and operation of a national CCS infrastructure.

There are open questions about what it will take to create a national CCS infrastructure. We need a thorough assessment of our nation's geologic CO_2 storage capacity and a critical appraisal of the pipeline network required and the issues of transporting carbon dioxide from its sources to storage sites. Even though short-haul carbon dioxide pipelines already exist in the U.S. for the purposes of enhanced oil recovery—we've been employing these techniques in my state of Colorado for more than thirty years—a more expansive carbon dioxide pipeline network clearly raises new issues about pipeline network requirements and regulation, regulatory classification of carbon dioxide, and pipeline safety.

The DOE through its Carbon Sequestration Regional Partnerships, the DOT through the independent U.S. Surface Transportation Board with regulatory jurisdiction for transporting carbon dioxide, the FERC with its experience in the regulation of natural gas and oil pipelines, and the EPA through its underground injection control program have the necessary expertise to assess the important issues dealing with carbon dioxide pipelines that would be needed to handle large-scale carbon sequestration in this country.

We introduced this pipeline study bill because there has been a void at the federal level in the attention given to the infrastructure needed to bring CCS to fruition. We believe your agencies have the regulatory authority to begin such a feasibility study now, but I am concerned by the lack of coordinated federal action to answer these fundamental questions. I look forward to having a frank discussion regarding our path forward.

Thank you, Mr. Chairman.

STATEMENT OF HON. JOHN KERRY, U.S. SENATOR FROM MASSACHUSETTS

Senator Kerry. Thanks, Mr. Chairman, Senator Domenici and the rest of the Senators on the committee. Thank you very, very much for giving us an opportunity just to share a few thoughts with you. I particularly appreciate the opportunity to talk about S. 2323 which is a bill that Senator Stevens and I have jointly introduced and I'll say a word about it in a moment.

But I just want to remind the committee of the underlying importance of what drives both pieces of legislation and our being here today. We all know that last year the Nobel Prize winning intergovernmental panel on climate change issued its latest and most comprehensive report reflecting the consensus of over 2,000 of the world's most respected climate scientists. That report established beyond any reasonable doubt the urgency of acting to address climate change.

I had the privilege of representing the Senate for a brief 36 hours, because of our votes at the end of the year, in Bali at the climate change negotiations. I must say I've been attending those conferences since 1992 when Al Gore, Tim Wirth and a bunch of us went down to Rio to the Earth Summit. I've met with the various delegations over the course of time including the Chinese.

This time I found the Chinese transformed, engaged, prepared to discuss how to measure what they do, obviously not quite at the same rate and same scale. It was an entirely different conversation than any that we have had yet and opens the door to what really needs to be done because China will be at our levels of emissions within 10 years. So, obviously, we're going to have to find a way to achieve this. But this is part of that mosaic, if you will, Mr. Chairman.

The science shows that—and I also found there a sense of urgency among finance ministers, prime ministers, foreign ministers, environment ministers, trade ministers, presidents, an unbelievable sense of urgency about this issue. The science also shows that coal combustion is one of the greatest contributors to climate change. Those of us that seek to deal with this issue understand we're going to have to deal with this component of it.

Coal is not going to go away in the near term, no matter how much we wish that in terms of its negative impact. Not the positive, but the negative. It's cheap. It's abundant here in America. Countries such as China are using it extensively. They're building approximately one coal fired plant per week, pulverized coal fired plant, without modern technology right now. Coal accounts for 80 percent of their CO_2 emissions.

So they are building infrastructure that's effectively going to pollute for years to come. We will get the results of that pollution because it blows over us and falls in rain and so forth. Frankly it's my judgment, I think the judgment of a lot of people that the international community needs to be far more concerned about this and urgent about this than it is.

That's why we have to rapidly develop and implement carbon capture and storage technology, which is the purpose of this hearing. It was recommended last year in a similar report by the Massachusetts Institute of Technology. This technology will enable us, providing it works according to all of the designs and ways in which they believe it will, to capture the emissions from power plants and other industrial facilities and permanently bury them in deep saline aquifers and other geological formations.

Two recent reports identified carbon capture and sequestration as the most promising area for emission reductions in the electric power sector. A December 2007 McKinsey study determined that by 2030, 9 percent of U.S. electricity could come from coal plants equipped with CCS. The Electric Power Research Institute, the research arm of the electric power industry, estimated this number at 15 percent.

I might say that I hope those figures reflect a growth without the level of intervention that ought to take place because if it isn't, it isn't going to get the job done. All of us need to understand that. If you believe the scientists and you heed their warnings and you have to keep the climate change to a two degree centigrade level and 450 parts per million of greenhouse gases. There is no way to achieve that at that level of coal fired growth. So we have an enormous challenge ahead of us.

These studies demonstrate the potential however for the application of CCS. The purpose behind our bill and I think Senator Coleman's bill is to accelerate this effort so we can let the marketplace decide what works. We're not going to pick a winner or loser. We want to get the technology out there. Let the marketplace decide which technology in fact works the best and most effectively.

Now the energy bill that you passed—that we passed in the Senate last summer is a great start. I extend my gratitude to this committee for the provisions to inventory the sequestration capacity and to conduct essential demonstration projects. The legislation Senator Stevens and I have introduced which is the Carbon Capture and Storage Technology Act of 2007 would establish three to five commercial scale sequestration facilities and three to five coal fired demonstration plants with carbon capture.

Now there are benefits to these that are not the purpose of this hearing today so I won't go into those. But today's hearing is specifically, I gather, focused on one provision of the bill, which is the regulatory framework that must be established to oversee carbon capture and storage activities. The regulatory framework is as urgent as getting the technology out. Obviously they go hand in hand.

We have to make sure that we implement these projects correctly. We've never conducted sequestration here in the United States on the scale that we're contemplating. In fact only three sites in the entire world have projects of this magnitude.

First and foremost we need to guarantee the permanent storage of the CO_2 that we inject in the ground. CO_2 is naturally buoyant. When it's injected into the earth it will seek the earth's surface at all times. So, all of our aggressive efforts to develop CCS technology would be wasted if we don't make the right choices about where to inject the CO_2 to avoid leakage that releases the CO_2 back into the atmosphere.

Second, as we advance this technology we'll be making site specific decisions about appropriate sequestration locations. We need to ensure that these injection sites, whether in deep saline formations or oil or gas fields are safe, secure and permanent. We need to develop national siting guidelines that will provide confidence in the injectivity, capacity and effectiveness of storage sites. We need to develop consistent and reliable monitoring and verification protocols that will assist with site assessment and planning and baseline and operational monitoring to ensure that the CO_2 remains permanently sequestered. Finally we need an early warning system that will alert us to potential leakage or failure issues at these sites.

Now many of these elements are highly technical, but they are all essential to ensuring the success of this technology in addressing climate change and in providing companies, investors and the public with confidence that they're getting what they pay for when they invest in carbon credits associated with CCS. Siting monitoring and verification regulation are also necessary to provide us with certainty they're avoiding any potential harmful public health or environmental outcome. For example precautions have to ensure that CO_2 injection sites don't result in seepage into drinking water aquifers and the release of heavy metals.

As we think through the regulatory framework for CCS, we have to be mindful that any CO_2 leakage within a contained environment could result in additional health or safety risks if not done properly. So for that reason the regulatory scheme is going to be critical. It will also determine whether or not this is going to work, folks. That is going to determine, very significantly, what our options are with respect to global climate change. So the faster we get about this business and the faster we get the demonstration projects out there properly, the better we're going to be able public choices for the long term.

To resolve these issues I've developed a provision in this legislation that directs the key agencies, including EPA, DOE and Interior to create a task force to develop comprehensive regulations to address the issues of leakage, public safety and environmental protection. These regulations would establish the regulatory framework to oversee the entire CCS process in a comprehensive fashion linking the complicated mechanisms for capture, transport, injection and storage of CO_2. The task force is specifically directed to consult with the industry as well as the technical experts in developing these regulations. The involvement of these experts, who've been involved in large scale sequestration projects abroad or enhanced oil recovery, which many of you are familiar with.

We have used this effort to drive oil out and capture oil today. So we have it in certain scale. But we need to develop the ability for the regulatory scheme to govern this process. Many of those individuals, incidentally, are behind me here testifying today. I'm eager to learn about their input as to how we do this most appropriately. I look forward to working with the committee as we try to meet this urgent challenge. Thank you, Mr. Chairman.

[The prepared statement of Senator Kerry follows:]

Prepared Statement of Hon. John Kerry, U.S. Senator from Massachusetts

Chairman Bingaman, Senator Domenici and colleagues—thank you for inviting me to testify today. I appreciate the opportunity to introduce an issue and a piece of legislation that I believe are critical to our efforts to combat global climate change.

Last year, the Nobel Prize-winning Intergovernmental Panel on Climate Changeissued its latest and most comprehensive report, reflecting the consensus of over 2,000 of the world's most respected climate scientists. The report established beyond any real doubt the urgency of acting to address climate change.

In the last 250 years, carbon dioxide levels in the atmosphere have risen from 280 parts per million to 380—higher today than at any time in the past 650,000 years. Scientists tell us that we have to keep CO_2 concentration below 450 parts per million—which corresponds to an increase of 2 degrees Celsius—to avoid a large scale catastrophe. And we only have ten years in which to act. But unless we take dramatic action, we're expected to reach 600–700 parts per million by the year 2100. This is urgent. It is being driven by facts and by the alarms that scientists across the planet are sounding today.

We who seek to fight climate change must face the reality that, in the foreseeable future, coal isn't going away. It's cheap and abundant here in America and in places like China, which is growing at 11% a year and building one coal-fired power plant per week. Today coal accounts for 80% of China's CO_2 emissions, and they and others are building infrastructure that will pollute for decades to come.

That is why it is critical that we run, not walk, to develop and implement carbon capture and storage technology, as recommended last year in a seminal report bythe Massachusetts Institute of Technology. This technology will enable us to capture the emissions from power plants and other industrial facilities, and permanently bury them in deep saline aquifers and other geological formations.

In fact, two recent reports identified CCS as the most promising area for emission reductions in the electric power sector. A December 2007 McKinsey study determined that, by 2030, 9% of US electricity could come from coal plants equipped with CCS. The Electric Power Research Institute, the research arm of the electric power industry, estimated this number at 15%. These studies demonstrate the tremendous potential for the application of CCS. Our government should be making significant commitments to advancing this technology.

The Energy Bill was a very good start—and I would like to extend my thanks to this committee for its leadership on key provisions to inventory our country's sequestration capacity and conduct essential demonstration projects.

In addition, I have introduced legislation with Senator Stevens—the Carbon Capture and Storage Technology Act of 2007—which would establish 3–5 commercial-scale sequestration facilities and 3–5 coal-fired demonstration plants with carbon capture.

I would be happy to discuss the benefits of these projects, but today's hearing isfocused on another component of the bill—the regulatory framework that we need to put in place to oversee carbon capture and sequestration activities. My bill establishes an interagency task force, chaired by the Administrator of the EPA, to develop regulations governing the complicated mechanisms and requirements for the capture, transport, injection and storage of carbon dioxide. The task force is specifically directed to consult with industry, as well as technical and legal experts, in developing these regulations—and the individuals who will be testifying this morning are some of the leading authorities in the country on these issues. I am eager to hear their thoughts.

I look forward to continuing to work with my colleagues to advance carbon capture and storage technology, and I thank you again for the opportunity to testify this morning.

The Chairman. Thank you very much.

Senator Coleman.

STATEMENT OF HON. NORM COLEMAN, U.S. SENATOR FROM MINNESOTA

Senator Coleman. Thank you, Mr. Chairman. It's a pleasure to be sitting by the side of my colleague Senator Kerry. Both of our approaches here proceed with a firm belief that it's important to get the technology out there. I firmly believe in it.

Thinking about the Chinese experience and what they're doing. The country that gets the technology out there, I think, is going to dominate the 21st century on economic terms. The Chinese are going to have to buy it. They're choking to death they're going to have to buy it from us. So we have this, I think, huge incentive to move forward and you have to have a framework for that incentive.

When I was a young person I dreamed of being a basketball player. My heroes were guys like Bob Cousy, Oscar Robertson, Earl the Pearl Monroe. I'm dating myself here by the way. That all ended when a coach told me, Coleman, you may be small, but you can't jump. It is bad when you have two reinforcing problems.

Our Nation has that. We are highly dependent on foreign sources of energy and we produce dangerous amounts of greenhouse gases. How do we solve one problem without exacerbating the other?

Mr. Chairman this committee under your leadership and that of the ranking member has boldly moved to address both. You've crafted two landmark pieces of legislation in the past several years: the Energy Independence Security Act of 2007 and the Energy Policy Act of 2005. These comprehensive bills address numerous critical energy and environmental challenges facing our Nation and they establish a firm foundation on which to build our Nation's energy future.

I firmly believe that a big part of that future is going to require figuring out how to utilize America's 250 year supply of coal in an environmentally friendly manner. By taking CO_2 produced in coal power plants and piping that CO_2 to a location where it can be permanently stored, I believe we can greatly add to the country's economic and even national security. That's why I've introduced the CO_2 Pipeline Study Act which is another step in this committee's efforts to address these issues in an informed and timely manner.

I want to thank a number of members of this committee who are original co-sponsors of the CO_2 Pipeline Study Act for their leadership. Senator Murkowski, who's here today, Senator Salazar, Landrieu, Johnson, Martinez and Bunning, your guidance and assistance were invaluable in drafting this legislation. The fact is we have an immense supply of coal available in this country. It is a relatively low cost energy source we do not need to import. Accordingly we do not need to send our valuable dollars overseas to hostile regimes in order to keep the lights on. We simply must find a way to use coal without jeopardizing the climate. Indeed coal already supplies about half our Nation's electric power.

The good news is my colleague Senator Kerry has testified about in greater detail is that we can take the CO_2 out of the emissions of a coal power plant and store it underground. More research needs to be done. But the future of CO_2 free coal plants looks bright. One of the key components of making CO_2 free coal is a reality of how to transport this gas from the power plant to the ground.

Currently there are many uncertainties about the rules and costs that will exist with the construction and operation of CO_2 pipelines. The CO_2 Pipeline Study Act will answer these questions. It will provide certainty to industry and to consumers. The CO_2 Pipeline Study Act seeks the input of a number of Federal agencies and departments: the Department of Energy, Interior, Transportation, the Federal Energy Regulatory Commission, FERC, the Environmental Protection Agency.

Each of these has expertise about a variety of issues associated with the building of pipelines.

These agencies are required to conduct the study of any technical, siting, financing or regulatory barriers that might prevent or impede the development of a carbon dioxide pipeline industry.

They're also asked to address any safety and integrity issues associated with constructing carbon dioxide pipelines. I anticipate the recommendations in their study may well serve as a basis of future congressional action on these issues.

In short, this bill will lay out the groundwork for CO_2 free coal plants that will allow America to move forward quickly, but carefully and responsibly to its piping CO_2. The CO_2 Pipeline Study Act also works in tandem with and complements the actions on that broader carbon dioxide issue taken in S. 2323, Senator Kerry's bill, also the Energy Independence and Securities Act of 2007 and the Energy Policy Act of 2005. These bills address carbon capture at the point of creation, for example at a coal fired power plant and the storage of carbon dioxide at an appropriate geologic formation.

However unless the coal fired plant happens to be near a suitable storage location, the carbon dioxide will have to be piped to an appropriate geologic formation to sequestration. That is what the Pipeline Study Act answers. It addresses the issues associated with transporting carbon dioxide from its point of capture to its point of storage for use in enhanced coal recovery.

We have an enormous potential domestic supply of energy. It can be used to cool and heat our homes, power our businesses and industries and create enumerable new jobs. However, our Nation will only realize these benefits if we can produce and use this energy in an environmentally sensitive manner. The CO_2 Pipeline Study Act is an important step in our efforts to develop this energy resource in an environmentally responsible way. We need to have the regulatory framework in place if we are going to get the technology out in time. Senator Kerry's bill and my bill begin that necessary and important conversation.

Chairman Bingaman, Ranking Member Domenici and members of this committee, I thank you for this opportunity to speak on behalf of S. 2144, the CO_2 Pipeline Study Act. With your leadership we are turning a national dilemma throwing energy dependence and greenhouse gas production into a win-win with the help of our people, our economy and ultimately our national security. Thank you, Mr. Chairman.

[The prepared statement of Senator Coleman follows:]

Prepared Statement of Hon. Norm Coleman, U.S. Senator from Minnesota

First, I want to thank Chairman Bingaman and Ranking Member Domenici for holding this important hearing today and inviting me to speak on behalf of the Carbon Dioxide Pipeline Study Act.

When I was a young person I dreamed of being a basketball player like Bob Cousy or Earl The Pearl Monroe. That all ended when a coach told me, ''Coleman, you may be small, but you can't jump.'' It's bad when you have two reinforcing problems.

Our nation has that. We are highly dependent on foreign sources of energy and we produce dangerous amounts of greenhouse gases. How do we solve one problem without exacerbating the other?

This committee, under your leadership, has boldly moved to address both. You have crafted two landmark pieces of legislation in the past several years: the Energy Independence and Security Act of 2007 and the Energy Policy Act of 2005. These comprehensive bills address numerous critical energy and environmental challenges facing our nation, and they establish a firm foundation on which to build our nation's energy future.

I firmly believe that a big part of that future is going to require figuring out how to utilize America's 250 year supply of coal in an environmentally friendly manner. By taking CO_2 produced in coal power plants and piping that CO_2 to a location where it can be permanently stored, I believe we can greatly add to the country's economic and even national security. That's why I've introduced the CO_2 Pipeline Study Act, which is another step in this committee's efforts to address these issues in an informed and timely manner.

I want to thank a number of Members of this committee who are original co-sponsors of the CO_2 Pipeline Study Act for their leadership: Senators Salazar, Murkowski, Landrieu, Johnson, Martinez and Bunning. Your guidance and assistance were invaluable in drafting this legislation.

The fact is, we have an immense supply of coal available in this country—it's an energy source we do not need to import, and accordingly, we do not need to send our valuable dollars overseas to hostile regimes in order to keep the lights on.

We simply must find a way use coal without jeopardizing the climate. Indeed, coal already supplies about half of our nation's electric power. The good news, as my colleague Senator Kerry has testified about in greater detail, is that we can take the CO_2 out of the emissions of a coal power plant and we can store it underground. More research needs to be done, but the future of CO_2-free coal plants looks bright. Yet one of the key components of making CO_2-free coal a reality is how to transport this gas from the power plant to the ground.

Currently, there are many uncertainties about the rules and costs that will exist for the construction and operation of CO_2 pipelines. The CO_2 Pipeline Study Act will answer these questions, it will provide certainty to industry and consumers.

The CO_2 Pipeline Study Act seeks the input of a number of federal agencies and departments—the Departments of Energy, Interior, and Transportation, the Federal Energy Regulatory Commission (FERC) and the Environmental Protection Agency (EPA). Each of these has expertise about a variety of issues associated with the building of pipelines.

The agencies are required to conduct a study of any technical, siting, financing, or regulatory barriers that might prevent or impede the development of a carbon dioxide pipeline industry. They are also asked to address any safety and integrity issues associated with constructing carbon dioxide pipelines. I anticipate the recommendations in their study may serve as a basis for future Congressional action on these issues. In short, this bill will lay the groundwork for CO_2-free coal plants, it will allow America to move forward quickly, but also carefully and responsibly toward piping CO_2.

The CO_2 Pipeline Study Act works in tandem with and compliments the actions on the broader carbon dioxide issue taken in S. 2323, the Carbon Capture and Storage Technology Act, the Energy Independence and Security Act of 2007 and the Energy Policy Act of 2005. These bills address carbon capture at the point of creation— for example at a coal fired power plant—and the storage of carbon dioxide at an appropriate geologic formation.

However, unless the coal fired power plant happens to be near a suitable storage location, the carbon dioxide will have to be ''piped'' to an appropriate geologic formation for sequestration. This is what the CO_2 Pipeline Study Act answers. It addresses the issues associated with transporting carbon dioxide from its point of capture to its point of storage or for use in enhanced oil recovery.

We have an enormous potential domestic supply of energy. It can be used to cool and heat our homes, power our businesses and industries and create innumerable new jobs. However, our nation will only realize these benefits if it can be produce and use this energy in an environmentally sensitive manner.

The CO_2 Pipeline Study Act is an important step in our efforts to develop this energy resource in an environmentally responsible way.

Chairman Bingaman, Ranking Member Domenici and members of this committee—thank you for this opportunity to speak on behalf of S. 2144, the CO_2 Pipeline Study Act.

With your leadership you are turning a national dilemma—growing energy dependence and greenhouse gases production—into a ''win-win'' for the health of our people, our economy and ultimately our national security.

Chairman. Thank you both very much. I think you made a contribution by the introduction of these bills and the efforts you've put into educating us on them.

Let me now either dismiss these witnesses, unless anybody has a question that's burning that they want to ask. We will allow you folks to get on with your other activities. But thank you again for being here.

We have two panels. Let me turn to Senator Domenici to make his opening statement and then I will call the first panel forward.

STATEMENT OF HON. PETE V. DOMENICI, U.S. SENATOR FROM NEW MEXICO

Senator Domenici. Thank you very much, Mr. Chairman. Good afternoon. I want to thank Senator Bingaman for scheduling this hearing and the witnesses for appearing. I'd like to thank Senators Coleman and Kerry for their work in drafting the measures before us.

Carbon sequestration, Mr. Chairman, holds real promise for reducing the emissions of greenhouse gases. Today however, that promise is far from realized. The technology has not been commercialized and a massive investment in infrastructure, pipelines, etc. is needed. As a result carbon sequestration must be viewed for what it is; a small piece of the solution to what is a larger issue of global climate change.

The recently passed energy bill included many provisions on this subject. It is recognized that an appropriate Federal role exists for researching, developing and commercializing cleaner technologies. It will be one thing to implement the Federal laws that we have passed, but we must also remember the economic law of diminishing returns.

Carbon sequestration as we know it is a classic example of that concept. The more aggressively we pursue it, the more it will cost. Because climate change is very much a global challenge, the benefits we derive will be incrementally smaller. America can be a leader in carbon sequestration. We have experience in the form of enhanced oil recovery to guide our investment and regulatory decisions.

Yes, if the United States acts unilaterally to reduce its emissions we risk saddling taxpayers with a steep price for minimal results. Other nations are on the verge of passing the United States in annual greenhouse gas emissions on a per capita basis, some already have. As greenhouse gas emissions decrease here at home increases in developing countries will more than offset our progress. Discussions of the carbon sequestration process are worthy of our committee's time.

I'll keep an open mind. I hope to learn what more can be done. But I also urge that my colleagues not put the cart before the horse. While we can and should advance this promising concept, we must know for sure that other countries will join us in this effort. I look forward to hearing from the witnesses that we have scheduled. I thank you again, Mr. Chairman.

The Chairman. Thank you very much. Let me call the first panel forward. First we have Chairman Kelliher from the Federal Energy Regulatory Commission. Benjamin Grumbles from the EPA. James Slutz, who is the Deputy Assistant Secretary in the Office of Oil and Natural Gas in the Department of Energy. Krista Edwards, Deputy Administrator with the Pipeline and Hazardous Materials Safety Administration in the Department of Transportation. Stephen Allred who is the Assistant Secretary for Land and Minerals Management in the Department of the Interior.

Thank you all very much for being here. If you could each take 5 to 6 minutes and just summarize the main points of your testimony for us, we will include your full testimony in the record, but I'm sure there will be questions of all of you.

Let me start with Chairman Kelliher and then Krista Edwards, Mr. Grumbles, Steve Allred and Mr. Slutz. So go right ahead Mr. Chairman. Thank you for being here.

Statement of Joseph T. Kelliher, Chairman, Federal Energy Regulatory Commission

Mr. Kelliher. Thank you, Mr. Chairman. Let me say first of all, thank you for the opportunity to be here today and the possibility of being here. My term of office would have expired at the end of last session. But in the waning minutes of the session I was confirmed along with my colleague Jon Wellinghoff. It reminded me a little bit like the Georgetown/West Virginia game the other night except the shot wasn't blocked at the last second. So I'm grateful for the support of the chairman, Senator Domenici, committee members and the staff.

Mr. Chairman, members of the committee, thank you for the opportunity to speak with you today about the regulatory aspects of carbon capture, transportation, sequestration and related legislation. My written testimony offers comments on the legislation before the committee and but in particular I want to say that FERC would be pleased to participate in the study required by S. 2144. We believe we can contribute to an examination of the regulatory barriers and regulatory options relating to the construction and operation of carbon dioxide pipelines.

I'm going to focus my oral testimony on the regulatory aspects of carbon transportation, the area where FERC's experience regarding pipelines may have the most value to the committee. While there are questions about carbon capture and sequestration technology, carbon dioxide transportation has been proven and storage of carbon dioxide has taken place for years. A network of carbon dioxide pipelines has been developed, mostly since the 1980s to promote enhanced oil recovery in declining oil fields. There is also some experience with storage of carbon dioxide.

Now up to this point the injection of carbon dioxide into oil production reservoirs has been a means of increasing oil production rather than an end unto itself. Storage takes place in the oil production fields rather than in reservoirs dedicated to carbon dioxide sequestration. Construction of the U.S. carbon dioxide pipeline network began over 25 years ago and that network now spans more than 3,900 miles. Siting of carbon dioxide pipelines has been governed by State law and to my knowledge state siting has worked well.

Under current law there is no Federal role in siting carbon dioxide pipelines. While operators of interstate carbon dioxide pipelines are free to set their own rates in terms of service, the U.S. Department of Transportation's Surface Transportation Board may hold proceedings to determine that rates are reasonable if a third party files a complaint. U.S. Department of Transportation's Office of Pipeline Safety within the Pipelines and Hazardous Material Safety Administration administers safety regulations governing interstate carbon dioxide pipelines.

The committee expressed an interest in exploring the regulatory aspects of carbon dioxide transportation. FERC is an infrastructure agency with nearly 90 years of experience regulating a broad range of energy infrastructure projects including oil and natural gas pipelines and related facilities. The United States has used three different regulatory schemes for pipeline transportation that might be relevant to congressional consideration of the regulatory aspects of carbon dioxide transportation.

First there is the model that has governed the existing carbon dioxide pipeline network. Under this approach pipelines are sited under State law. Transportation rates are set by the Surface Transportation Board when a complaint filed regarding rates is filed. The Office of Pipeline Safety ensures safety.

Second there is the oil pipeline model. Under this model oil pipelines are sited under State law. FERC sets the transportation rate. FERC has no siting or safety role with safety issues being handled by the Department of Transportation Office of Pipeline Safety. This model also has worked well for decades.

The third model is the natural gas pipeline model. Under the current version of this model FERC both sites interstate natural gas pipelines and sets their transportation rates. It may be useful to note however, that the original version of the Natural Gas Act, the 1938 Act, provided for state siting of interstate natural gas pipelines.

In 1947 however, Congress reached the conclusion that State siting of natural gas pipelines had failed and that it was necessary to resort to Federal siting. Congress amended the Natural Gas Act and provided for exclusive and preemptive Federal siting of interstate natural gas pipelines. While the Commission, while FERC is responsible for safety issues during the siting and construction phases, safety jurisdiction shifts to the Department of Transportation through the Pipeline and Hazardous Materials Safety Administration once construction is complete.

Now in my view any of these three approaches could prove effective in regulating a network of carbon dioxide pipelines. I have no reason to believe the existing regulatory scheme administered by the Surface Transportation Board is inadequate. In particular I would not recommend that Congress preempt the states in siting carbon dioxide pipelines by providing for exclusive and preemptive Federal siting. The precondition that led Congress to such a course for siting natural gas pipelines, the failure of State siting, does not appear to exist here. Further I would not recommend that Congress alter the safety role of the Pipeline and Hazardous Materials Safety Administration.

I appreciate the opportunity to testify here today and would be pleased to answer any questions you might have. Thank you.

[The prepared statement of Mr. Kelliher follows:]

Prepared Statement of Joseph T. Kelliher, Chairman, Federal Energy Regulatory Commission

Mr. Chairman and members of the committee, thank you for the opportunity to speak with you today about the regulatory aspects of carbon capture, transportation, and sequestration and two related bills, namely S. 2144, the

''Carbon Dioxide Pipeline Study Act of 2007'', and S. 2323, the ''Carbon Capture and Storage Technology Act of 2007''. I commend the committee for holding this hearing.

Carbon Capture and Sequestration Technology

Development of carbon capture and sequestration technology is an importantneed. There are questions about carbon capture and sequestration technology. The two bills that are the subject of this hearing address this need by requiring studies and funding research and development and demonstration projects. If these efforts are successful, carbon capture and sequestration may become a practical reality.

S. 2144

S. 2144 would direct the Secretary of Energy, in coordination with the Federal Energy Regulatory Commission (FERC), the Secretary of Transportation, the Administrator of the U.S. Environmental Protection Agency, and the Secretary of the Interior, to conduct a study to assess the feasibility of the construction and operation of pipelines to be used for the transportation of carbon dioxide for the purpose of sequestration or enhanced oil recovery and carbon dioxide sequestration facilities.

FERC has extensive experience in the siting and regulation of a wide variety ofenergy infrastructure projects, and we would be pleased to participate in the study required by S. 2144. In particular, FERC can play a helpful role examining regulatory barriers and regulatory options relating to the construction and operation of carbon dioxide pipelines, as provided by section 2(b) of the bill.

S. 2323

As I indicated above, there are questions relating to carbon capture and sequestration technology. This bill would address those questions directly, by funding carbon dioxide capture and storage research and development, and both carbon dioxide capture and sequestration demonstration projects. The bill has other provisions relating to establishment of an interagency task force to develop regulations for carbon dioxide capture and storage, an assessment of carbon dioxide storage capacity, and technology agreements.

Regulatory Aspects of Carbon Dioxide Transportation

While there are questions about carbon capture and sequestration technology, carbon dioxide transportation has been proven and storage of carbon dioxide has taken place for years. A network of carbon dioxide pipelines has been developed, mostlysince the 1980s, to promote enhanced oil recovery in declining oil fields. There is also some experience with storage of carbon dioxide.

Up to this point, the injection of carbon dioxide into oil production reservoirs has been a means of increasing oil production, rather than an end unto itself. Storage takes place in the oil production fields themselves, rather than in reservoirs dedicated to carbon dioxide sequestration. Enhanced oil recovery results in the storageof carbon dioxide in depleted production reservoirs.

I am not aware of whether any information has been developed regarding the leakage of carbon dioxide from the existing pipeline network or production fields. This might be an area worthy of research and development.

Besides enhanced oil recovery, carbon dioxide has been used for other purposes, including refrigeration and cooling, casting metal molds, welding, sandblasting,methanol and urea production, carbonation, and medical purposes.

Construction of the U.S. carbon dioxide pipeline network began over 25 years ago, and that network now spans more than 3,900 miles. Siting of carbon dioxide pipelines has been governed by state law, and to my knowledge state siting has worked well. Under current law, there is no federal role in siting carbon dioxide pipelines. While operators of interstate carbon dioxide pipelines are free to set their own ratesand terms of service, the U.S. Department of Transportation's Surface Transportation Board may hold proceedings to determine that rates are reasonable, but only if a third party files a complaint. Under the Interstate Commerce Termination Act of 1995, the Surface Transportation Board regulates interstate pipelines transporting commodities other than water, oil, or natural gas. The U.S. Department of Transportation's Office of Pipeline Safety, within the Pipelines and Hazardous Materials Safety Administration (PHMSA), administers safety regulations governing interstate carbon dioxide pipelines.

The committee expressed an interest in exploring the regulatory aspects of carbon dioxide transportation. FERC has a great deal of experience regulating energy infrastructure. The original mission of the agency was development of energy infrastructure, specifically licensing and regulating non-federal hydropower projects. Our infrastructure role has expanded over time to include natural gas pipelines and associated facilities, oil pipelines, and more recently we have been given a limited role in electric transmission siting.

The U.S. has used three different regulatory schemes for transportation of energy resources by pipeline that might be relevant to Congressional consideration of the regulatory aspects of carbon dioxide transportation. First, there is the model that has governed the existing carbon dioxide pipeline network, namely continuing the current regulatory scheme for interstate carbon dioxide pipelines. Under this approach, pipelines are sited under state law, transportation rates are set by the Surface Transportation Board when a complaint regarding rates is filed, and the Office of Pipeline Safety ensures safety.

Second, there is the oil pipeline model. Under this model, oil pipelines are sited under state law and FERC sets the transportation rate. FERC has no siting role or safety role (safety issues being handled by the Department of Transportation). This model has worked well for decades.

The third model is the natural gas pipeline model. Under the current version of this model, FERC both sites interstate natural gas pipelines and sets their transportation rates. It may be useful to note that the original version of the 1933 Natural Gas Act provided for state siting of interstate natural gas pipelines. However, in 1947 Congress reached the conclusion that state siting of natural gas pipelines had failed, and it was necessary to resort to federal siting. Congress amended the Natural Gas Act and provided for exclusive and preemptive federal siting of interstate natural gas pipelines. While the Commission is responsible for safety issues during the siting and construction phases, safety jurisdiction shifts to the Department of Transportation, though PHMSA, once construction is complete.

In my view, any of these three approaches could prove effective in overseeing a network of carbon dioxide pipelines. I have no reason to believe the existing regulatory scheme administered by the Surface Transportation Board is inadequate. In particular, I would not recommend that Congress preempt the states on siting carbon dioxide pipelines, by providing for exclusive and preemptive federal siting of carbon dioxide pipelines. The precondition that led Congress to such a course for siting natural gas pipelines—the failure of state siting—does not exist here. Further, I would not recommend that Congress alter PHMSA's safety role.

I appreciate the opportunity to testify before you today and would be pleased to answer any questions you may have.

Senator Domenici [presiding]: Ms. Edwards, you're next, ma'am.

STATEMENT OF KRISTA L. EDWARDS, DEPUTY ADMINISTRATOR, PIPELINE AND HAZARDOUS MATERIALS SAFETY ADMINISTRATION, DEPARTMENT OF TRANSPORTATION

Ms. Edwards. Ah, yes. Ranking Member Domenici, members of the committee, I appreciate the opportunity to discuss the safety programs administered by the Department of Transportation's Pipeline and Hazardous Materials Safety Administration and PHMSA's role in overseeing the safe transportation of carbon dioxide.

First on behalf of Secretary Peters I want to express PHMSA's strong support for the committee's efforts. We share your commitment to energy security and environmental protection. We are pleased to have a seat at the table as the committee considers the transportation requirements associated with carbon capture and storage.

DOT has vast experience managing the risks of CO_2 in all modes of transportation. Under the hazardous materials transportation law the Department has long overseen the movement of CO_2 by rail, highway, air and vessel. PHMSA's hazardous materials regulations established standards for the design, testing and filling of tanks and other packages used to contain and store CO_2 in each of its physical states as a gas, liquid and solid.

Since 1991, DOT has also overseen the transportation of CO_2 by pipeline. Together with our State partners, PHMSA currently oversees the operation of nearly 4,000 miles of CO_2 pipelines. We expect that that number will grow as the Nation moves ahead with carbon capture and sequestration. So I am pleased to assure the committee that DOT existing pipeline safety program will fit new CO_2 pipelines however they may be configured.

Congress reauthorized the program only a little more than a year ago, we thank you for that, reflecting strong support for PHMSA's mission and approach. Our approach is performance based. We aim to protect people, property and the environment by preventing pipeline failures and by mitigating the consequences of those that occur. Our integrity management programs promote continuous improvement by requiring operators to identify all threats to pipeline integrity and to remedy safety problems in priority order, worse first. By identifying and reducing defects before they grow to failure our integrity management programs are driving significant improvement in safety performance and reliability.

Our national pipeline safety program provides seamless oversight of pipeline operations through PHMSA's five regional offices and strong partnerships with our pipeline safety programs. The State programs play a critical role directly overseeing the vast share of pipeline infrastructure including most intrastate pipelines.

To meet our goals PHMSA also must be more than a regulator. We are supporting the development of new technologies such as tools for improving assessment of pipelines and non-regulatory initiatives such as the nationwide campaign to promote safe excavation practices. We work with all stakeholders who can contribute to safety outcomes including communities near new, existing and planned pipelines. As part of a comprehensive approach to pipeline safety we believe in preparing communities to make risk informed land use decisions and in preparing local first responders to respond to pipeline incidents.

Although PHMSA has no authority over pipeline siting we work closely with FERC and DOI in reviewing designs for pipelines and in responding to local concerns about pipeline safety. These efforts are paying off in terms of improved safety, reliability of supply and public confidence. The number of significant pipeline incidents has reached historic lows even as the size of the pipeline network has grown. Within these data I'm very pleased to report that the safety record for CO_2 pipelines is particularly good. There's been no loss of life and no injuries on DOT regulated CO_2 pipelines.

In closing I want to reiterate our strong support for the development of new energy solutions. PHMSA is pleased to work with the committee, our Federal and State partners and industry to prepare for the safe operation of new and expanded CO_2 pipelines. We offer our agency's expertise and experience as the committee considers and addresses future requirements for carbon capture and storage.

Thank you again for the opportunity to testify. I'll be pleased to answer any questions.

[The prepared statement of Ms. Edwards follows:]

Prepared Statement of Krista L. Edwards, Deputy Administrator, Pipeline and Hazardous Materials Safety Administration, Department of Transportation

Chairman Bingaman, Ranking Member Domenici, members of the committee: Thank you for the opportunity to discuss the safety programs administered by the U.S. Department of Transportation's Pipeline and Hazardous Materials Safety Administration (PHMSA) and our experience in overseeing the commercial transportation of carbon dioxide.

As the committee considers future requirements for carbon capture and sequestration, I am pleased to confirm that large volumes of carbon dioxide (CO_2) are shipped safely in the U.S. today, including by pipeline. PHMSA's existing programs andstandards governing CO_2 transportation provide effective protection to life and property, with due regard for the efficiency and performance of the transportation system.

As the DOT agency with jurisdiction over the movement of hazardous materials by all transportation modes, PHMSA has extensive experience managing the risks of compressed CO_2, in each of its physical states: gas, liquid, and solid (dry ice). Unlike natural gas and other gases regulated as hazardous materials, CO_2 is noncombustible and nontoxic. A colorless, odorless by-product of human respiration, CO_2 is present naturally in the environment and, at normal atmospheric levels, is vital to plant life and poses no immediate hazard to people or animals. In higher concentrations, as when CO_2 is contained for transport or storage, exposure to CO_2 can cause respiratory problems, including suffocation. CO_2 reaches its liquid stateat combinations of high pressure and low temperature. Both variables affect the consequence of a release of liquefied CO_2 in each case depending on the proximity of people and the location and surrounding conditions. In a remote, unpopulated area, even a large release of liquefied CO_2 will vaporize harmlessly into the atmosphere and is unlikely to cause serious injury. By contrast, a large, sudden release of liquefied CO_2 could have catastrophic consequences in a populated area. Becauseit is heavier than air, compressed CO_2 tends to pool near the ground, displacing all oxygen, and form a vapor cloud as it dissipates.

Because of these properties when compressed and/or in high concentrations, CO_2 has long been considered a hazardous material subject to the Hazardous Materials Transportation Laws, 49 U.S.C. 5101 et seq., and DOT's implementing regulations, 49 C.F.R. parts 171–180, governing transportation by air, rail, highway, and water.PHMSA's Hazardous Materials Regulations (HMR) prescribe a comprehensive risk management framework for CO_2 transport, covering classification, packaging, handling, and hazard communication (shipping documentation and labeling). The packaging standards for CO_2 transport vary based on volume, pressure, and transportation mode; in each case, the HMR mandate the use of an approved cylinder or tank, subject to specific requirements for design, testing, certification, and filling.

The Department assumed oversight of CO_2 pipelines in 1988, under legislation directing the Secretary to develop regulations for the safe transportation of CO_2 by pipeline. Pursuant to the mandate, in 1991, the Department extended its existing hazardous liquid pipeline rules (49 C.F.R. part

195) to these operations. CO_2 pipelines became subject to additional integrity management requirements when the liquid IM program was adopted in 2000.

As with liquid operations generally, PHMSA shares oversight of certain CO_2 pipelines with authorized State programs. Together with these State partners, PHMSA currently oversees close to 4,000 miles of CO_2 transmission pipelines (as depicted in *Figure 1)—amounting to roughly five percent of all hazardous liquid pipeline mileage under our jurisdiction. Of these CO_2 lines, approximately 66 percent are interstate (crossing State borders) pipelines with the remaining 34 percent classified as intrastate (within State borders). Located primarily in the States of Texas, New Mexico and Wyoming, these pipelines deliver CO_2 for a variety of industrial purposes, including enhanced oil recovery activities. Within the national pipeline network as a whole, the CO_2 lines are relatively new: approximately 91 percent were constructed after 1980.

As the Administration and Congress work to enhance our Nation's energy security and protect the environment, we understand the need to extend the transportation infrastructure—including the delivery of alternative fuels and the transport of CO_2 for sequestration or use in energy production. And we understand the importance of pipeline transportation for safe and efficient movement of large volumes of haz-ardous materials. With the right risk controls in place, pipelines can operate safely anywhere—it's not a matter of ''if,'' but ''how.''

PHMSA's pipeline safety program aims to promote continuous improvement in public safety, environmental protection, and system performance by identifying and addressing all threats to pipeline integrity and mitigating the consequences of pipe-line failures. Our regulations cover the design, construction, maintenance, and operation of liquefied natural gas (LNG) facilities and hazardous liquid and natural gas pipelines, both interstate and intrastate, including the gas distribution systems that directly serve homes and businesses. We work closely with national and international standards organizations and encourage the development of consensus standards complementing our performance-based regulations.

Our integrity management regulations, which currently apply to transmission pipelines (liquid and gas), require operators to conduct risk assessments of the condition of their pipelines; develop and implement risk control measures to remedy safety problems, worst first; and evaluate and report on program progress and effectiveness. Under integrity management programs, operators are identifying and repairing pipeline defects before they grow to failure, producing steady declines in the numbers of serious incidents.

Along with risk-based standards and practices, technological advances are driving significant improvement in the control of pipeline risks. PHMSA administers a cooperative research program that promotes the development of new methods, materials, and tools for improving leak detection systems and detecting and preventing corrosion, outside force damage, and other threats to pipeline integrity. We work closely with informed stakeholders, including other Federal agencies, our State partners, and industry, to target our limited R&D funding on promising technologies to address the most urgent safety issues. Most recently, in preparation for the growing use of alternative fuels, our R&D program is focused extensively on technical issues associated with the movement by pipeline of ethanol and ethanol-blended fuels.

As an agency dedicated to the safe transportation of hazardous materials, PHMSA must be more than a regulator. Our success depends on our ability to leverage non-regulatory solutions and to work closely with all stakeholders who can contribute to safety outcomes, including communities in the path of existing or new pipelines. Although PHMSA has no authority in pipeline siting, we work closely with the Federal Energy Regulatory Commission (FERC) in reviewing designs for proposed gas transmission pipelines and liquefied natural gas (LNG) facilities and in responding to local concerns about pipeline safety. We consult with other Federal and State agencies on how our regulatory requirements relate to their permitting decisions about pipelines. Recognizing that public decisions affecting transportation and energy supply often must be made at a national level, we believe a pipeline safety program can and must involve local communities, including zoning and planning officials and emergency responders. As part of a comprehensive approach to pipeline safety, we believe in preparing communities to make risk-informed land use decisions and in building local capability to respond to pipeline incidents. PHMSA works closely with fire service organizations on numerous safety projects, including the development of training standards and educational materials concerning pipeline incident response.

To carry out our oversight responsibilities, PHMSA operates five regional pipeline safety offices and is authorized to employ 111 inspection and enforcement professionals for fiscal year 2008. In addition to compliance monitoring and enforcement, PHMSA's regional offices respond to and investigate pipeline incidents and participate in the development of pipeline safety rules and technical standards. Our regional offices also work closely with PHMSA's State program partners, which employ approximately 400 pipeline inspectors and directly oversee the largest share of the U.S. pipeline network, including most intrastate pipelines. Under our Congressionally-authorized Community Assistance and Technical Services (CATS) program, PHMSA's regional offices provide safety-focused community outreach and education. With the current wave of pipeline expansion, and increasing commercial and residential development around existing pipelines, the CATS program is serving a vital role in educating the public about pipeline safety and encouraging risk informed land use planning and safe excavation practices.

With safety our top priority, under Secretary Peters' leadership, the Department is targeting the prevention of all transportation-related deaths and injuries. Although further improvement is needed, the safety record for hazardous materials transportation is good and getting better in all sectors, including hazardous liquid pipeline operations. Since the introduction of IM programs in 2000, the annual number of serious incidents involving hazardous liquid pipelines has reached historic lows, even as the size of the pipeline network has grown. Although the data sets are not yet large enough to make statistically significant comparisons, the trend line over the past 20 years (as depicted in *Figure 2) is favorable.

Within these data, the safety record for CO_2 pipelines is particularly good. Of the 3,695 serious accidents reported on hazardous liquid pipelines since 1994, only 36 involved CO_2 pipelines. Among the 36 incidents, only one injury, and no fatalities, was reported. In all other instances, the accidents were classified as serious based on the extent of property damage (including damage to the pipeline facility) or product loss.

With the benefit of this experience and record, PHMSA is pleased to work with the committee, our Federal and State partners, and industry to prepare for the safe operation of new or extended CO_2 pipelines. The existing pipeline safety program administered by PHMSA has provided effective oversight of CO_2 pipelines since 1991 and will accommodate new and expanded carbon dioxide pipelines, however they are configured. We are happy to work with the Department of Energy and other Federal partners to evaluate the feasibility of particular pipeline configurations and/or plan for their development.

Likewise, PHMSA is committed to working with any agency or agencies involved in siting CO_2 pipelines, just as we work with FERC today in connection with the licensing of gas transmission pipelines and LNG facilities. We offer our agency's considerable experience and technical expertise to the committee as it considers and addresses the transportation requirements associated with CO_2 capture and sequestration.

Mr. Chairman, I want to assure you and members of the committee that the Administration, Secretary Peters, and the dedicated men and women of PHMSA share your strong commitment to safe, clean, and reliable pipeline transportation. Like you, we understand the importance of PHMSA's mission to the Nation's economic prosperity and energy security, and we look forward to working with the committee to address the current challenges. Thank you.

Senator Domenici. Thank you very much. Senator, do you want me to proceed?

Senator Dorgan. Why don't you, the chairman will be here momentarily.

Senator Domenici. All right. Let me ask the two of you who have already testified, neither of you made a statement about quality of CO_2. Could I start with you, Mr. Chairman? Is CO_2 dangerous?

Mr. Kelliher. Is CO_2 dangerous?

Senator Domenici. I said is it dangerous like natural gas?

Mr. Kelliher. No, it's not like natural gas storage.

Senator Domenici. Does it blow up?

Mr. Kelliher. No, sir.

Senator Domenici. Will it hurt people?

Mr. Kelliher. Not in the way a natural gas leak could hurt people.

Senator Domenici. I imagine if you got too much you would go to sleep, right? Does it hurt people, ma'am?

Ms. Edwards. I'm pleased to speak to the hazard, thank you. We do regulate CO_2 in each of its forms as a hazardous material for the purposes of transportation. Certainly there are risks and hazards associated with it that's why it's under both of our safety programs.

On the other hand, as you pointed out, the hazards associated with this material are different and in many ways less than the hazards associated with other materials that are part of our—that we manage the oversight of through pipelines or significantly CO_2 in certain concentrations will cause respiratory problems in humans and could cause suffocation in a situation in which, you know, a human was exposed to intense concentrations. It tends— it's heavier than air as a liquid which is the form in which it's transported in pipelines so a massive release in a populated area without the right conditions and there are variables having to do with ventilation and temperature, of course, in terms of its vaporization and the rate of vaporization. But, you know, there have been, again I reiterate that our safety record for CO_2 pipelines has been very good. But, you know, it is why we oversee its movement in transportation because it poses a hazard.

Congress directed us to take on this oversight in 1988.

Senator Domenici. Thank you very much, ma'am.

Proceed, sir.

STATEMENT OF BENJAMIN H. GRUMBLES, ASSISTANT ADMINISTRATOR FOR WATER, ENVIRONMENTAL PROTECTION AGENCY

Mr. Grumbles. Thank you, Senator.

Senator Domenici. You're welcome.

Mr. Grumbles. Members of the committee I'm Benjamin Grumbles. I'm the Assistant Administrator for Water at the U.S. EPA. I really appreciate the opportunity to discuss EPA's important work on the regulatory aspects of carbon sequestration.

The Administration is committed to taking timely and responsible actions to confront the serious challenges of global climate change. EPA, in particular, believes innovative solutions will be critical to meeting this long term challenge including technologies and practices to mitigate greenhouse gas emissions. Carbon capture and storage is one of a portfolio of innovative technologies that could make a significant contribution to reducing greenhouse gas emissions. EPA is committed to advancing such efforts in a manner consistent with our obligations to safeguard public health and the environment as required by the Safe Drinking Water Act.

Carbon sequestration isn't a silver bullet, but it may be an ace in the hole for mitigating greenhouse gas emissions. We're very excited at EPA about the recent activities that the Administrator announced in October of last year that the Agency would move forward with a rulemaking under our authorities of the Safe Drinking Water Act. Our current schedule is an accelerated schedule. But our schedule is to propose regulations under the Safe Drinking Water Act's Underground Injection Control Program by the summer of this year.

The Agency is engaged in many efforts with our partners in Federal, State and local government and in the private sector on the wide array of carbon capture and storage and sequestration matters. But my testimony does focus on the regulatory aspects of injection and sequestration. Over the past several years we've been coordinating with the Department of Energy, the lead agency for research and development, and working with them on in support of their efforts on the carbon sequestration technology road map.

In March 2007, EPA issued technical guidance under our Safe Drinking Act authorities to help State and EPA regional managers in processing permit applications for experimental well permits for carbon sequestration. As I mentioned the Administrator then followed that up with an announcement in October 2007 that we are now fully committed to moving forward with a rulemaking, for full scale, not just experimental, but for full scale geo-sequestration of carbon dioxide recovered from emissions of coal fired power plants and other facilities. The proposed regulation which is currently in development will take into account our UIC program requirements that we have for the existing classes under the UIC program.

The key components of the proposed regulation will include requirements related to geologic site characterization to ensure wells are sited in suitable areas to limit the potential for migration of injected and formation fluids into an underground source of drinking water. The proposed regulation that we're working on will include well construction and operation requirements to ensure wells are properly constructed and managed. Mechanical integrity testing for the wells, monitoring for the wells and also, importantly, well closure, post closure care and also financial responsibility requirements regarding the proper plugging and abandonment of injection wells.

Importantly we will also be discussing long term liability and seek further comment on the issue as part of the proposed rule-making. We recognize there will need to be a robust debate on this important issue. We're expecting that once this rule is proposed that we will take next steps, coordinate with our Federal colleagues, review public comments. We're estimating a final rule in late 2010 or sometime in 2011.

The rule will embrace the concept of adaptive management. We're using an adaptive approach that will allow the agency to collect information and use data from DOE demonstration and other early projects to inform the final regulation and any subsequent revisions, if necessary. The hallmark of progress for us on this is continued coordination at the Federal level, but also at the State level with our State partners, whether it's IOGCC or the Ground Water Protection Council and at the local level and the national laboratories and with the private sector.

Mr. Chairman, I would just underscore the importance of this. It's one of the Administrator's priorities in our agency's own climate change clean energy strategy. I look forward to answering questions members of the committee might have. Thank you, Mr. Chairman.

[The prepared statement of Mr. Grumbles follows:]

Prepared Statement of Benjamin H. Grumbles, Assistant Administrator for Water, Environmental Protection Agency

Thank you, Chairman Bingaman and Members of the committee. I am Benjamin H. Grumbles, the Assistant Administrator for Water at the EPA, and I appreciate the opportunity to describe the Agency's important work on regulatory aspects of carbon dioxide sequestration.

This Administration is committed to taking timely and responsible actions to confront the serious challenge of global climate change. EPA believes innovative solutions will be critical to meeting this long-term challenge, including technologies and practices to mitigate greenhouse gas emissions. The Administration is actively investigating the prospects for carbon capture and storage (CCS), a process that involves capturing carbon dioxide (CO_2) from power plants and other industrial sources and injecting it into deep subsurface geologic formations for long-term storage. CCS is one of a portfolio of innovative technologies that could make a significant contribution to reducing greenhouse gas emissions to the atmosphere and EPA is committed to advancing such efforts in a manner consistent with our obligation to safeguard public health and the environment as required by the Safe Drinking Water Act (SDWA).

EPA staff are evaluating many aspects of CCS technology and deployment, focusing our efforts in two areas: (1) partnering with public and private stakeholders to develop an understanding of the environmental aspects of carbon capture and storage that must be managed for the necessary technologies to become a viable strategy for reducing greenhouse gases; and (2) ensuring carbon dioxide storage is conducted in a manner that protects underground sources of drinking water. My testimony focuses on the second of these two areas, EPA's development of a regulation for geologic sequestration (GS) of CO_2 and the collaboration taking place to support such efforts, all of which are relevant to your consideration of Section 5 of Senate Bill 2323.

Over the past several years, EPA has been coordinating with the Department of Energy (DOE), the lead agency for research and development of CCS technology. As DOE has developed a Carbon Sequestration Technology Roadmap for the development and deployment of this technology, EPA has been working to design an appropriate management framework for geologic sequestration. By engaging in DOE's expansive R&D program early and working with stakeholders on all sides of this issue, EPA is well positioned to help in the permitting of future carbon dioxide underground injection wells.

Regulatory Scope, Content, and Timeframe

EPA has statutory authority under the SDWA to carry out the Underground Injection Control (UIC) program to protect underground sources ofdrinking water from the injection of fluids for disposal or storage. In March 2007, EPA issued technical guidance to help State and EPA Regional UIC managers in processing permit applications for GS demonstration projects under the general UIC regulations. Recognizing that the technology is rapidly progressing towards full-scale deployment, Administrator Stephen Johnson announced, in October 2007, EPA's plans for developing national rules for full-scale GS of carbon dioxide recovered from emissions of coal-fired power plants and other facilities. EPA will propose regulations in the Federal Register this Summer to ensure that carbon dioxide injection is done in a manner that does not endanger underground sources of drinking water.

Under the SDWA, EPA develops minimum requirements for state UIC programs. States may develop their own regulations for injection wells in their State. These requirements must be at least as stringent as the federal requirements (and may be more stringent). Annually, billions of gallons of fluids are injected underground through wells authorized under State and Federal UIC Programs. This includes approximately 35 million tons of carbon dioxide that are injected for the purposes of enhancing oil and gas recovery. EPA's proposed regulations will build on the UIC Program's many years of experience in safely injecting fluids, including carbon dioxide, into the subsurface.

The proposed regulation, currently in development under an accelerated schedule, will take into account the EPA's existing UIC program requirements. Key components of the proposed regulation will include requirements related to: (1) geologic site characterization to ensure that wells are sited in suitable areas to limit the potential for migration of injected and formation fluids into an underground source of drinking water; (2) well construction and well operation to ensure that the wells are properly constructed and managed; (3) well integrity testing and monitoring to ensure that the wells perform as designed; and, (4) well closure, post-closure care and financial responsibility to ensure proper plugging and abandonment of the injection well. We will also discuss long-term liability and seek further comment on this issue as part of the proposed rulemaking.

Importantly, the proposal will also include public participation requirements that would be associated with issuance of permits. We will assess the costs of carrying out regulations for geologic sequestration programs as part of the economic analysis for the rulemaking.

EPA is reviewing available data on existing demonstration projects to inform our decision-making and development of the rule. Once a proposal is published, EPA will review public comments and take into account any new data and demonstration project outcomes prior to publishing a final rule by 2011. EPA's timeframe for the proposed rulemaking is consistent with the time frame for the DOE Roadmap, which projects fullscale project deployment to begin in the 2012–

2020 timeframe. To ensure that GS can be deployed as rapidly and safely as possible, EPA is using an adaptive approach that will allow the Agency to collect information and use data from DOE demonstration and other early projects to inform the final regulation and any subsequent revisions, if necessary.

Coordination and Collaboration

Within EPA, the Office of Water and Office of Air and Radiation are working together on all activities related to geologic sequestration in order to conduct technical and economic analyses, develop risk management strategies, collaborate with key stakeholders, and clarify the relationships among various statutes (including theSafe Drinking Water Act and Clean Air Act) and EPA regulations.

EPA is working closely with DOE to leverage existing efforts and technical expertise. EPA and DOE are coordinating with Lawrence Berkeley National Laboratory to answer key technical questions regarding impacts on groundwater and under-ground formations. The Agency is also monitoring the progress of research being conducted by organizations such as the Pacific Northwest National Laboratory, Lawrence Livermore National Laboratory, and international projects such as Sleipner, In Salah, and Weyburn to help inform the regulatory framework.

The DOE's Regional Carbon Sequestration Partnerships are conducting demonstration projects to gather data on the effectiveness and safety of GS. These Regional Partnerships will implement many small and large-scale field tests of carbon dioxide injection throughout the country in a variety of geologic settings. One goal of the technical permitting guidance EPA issued in March of 2007 is to promote the exchange of information to support the development of a long-term GS management strategy.

EPA will also engage with the Department of Transportation, Department of Interior, States, and Tribes during the rulemaking process. EPA has worked closely with key organizations such as the Groundwater Protection Council (GWPC) and the Interstate Oil and Gas Compact Commission (IOGCC), which represent States that implement UIC programs, and we will continue to do so throughout the regulatory process. For example, the Agency has reviewed the IOGCC report entitled "Storage of Carbon Dioxide in Geologic Structures: A Legal and Regulatory Guide for States and Provinces." The document's discussion of issues such as permitting and property rights may be very useful as we develop regulations.

In December 2007, EPA established a workgroup on geologic sequestration to provide input to the proposed regulation. The workgroup includes EPA and

DOE staff, as well as representatives of four state agencies, two of whom were recommended by the IOGCC and two by GWPC. Thus far, the workgroup has provided input on various aspects of the regulatory framework and has begun to draft issue papers on key issues.

Over the past several years, the Agency has been holding workshops, attending conferences and meeting with public and private stakeholders including industry experts, legal experts, technical experts, and environmental advocates to gather useful input. Our past experience gives us confidence we can work closely with key stake-holders and experts to develop well-designed regulatory approaches.

This past December, EPA held a meeting that focused on the potential regulatory framework for geologic sequestration. The two day workshop, held in Washington, DC, was attended by more than 200 stakeholders representing government, research institutions, industry, public interest groups, law firms, and the general public. Another stakeholder meeting is planned for February 26 and 27, 2008 in Alexandria, Virginia. Additionally, over the past year EPA has held technical workshops with researchers and stakeholders to discuss technical considerations for establishing a GS framework.

CONCLUSION

EPA is committed to working with our public and private partners to accelerate the important work underway to realize the significant potential of carbon dioxide capture and geologic storage. EPA will continue to engage with other federal agencies and encourage participation of states, associations, public interest groups, industry, and other stakeholders as the Agency moves forward on this critical path towards development of a regulatory framework. Consistent with the goal of Senator Kerry's bill, our goal is to develop sound regulations that will enable full-scale CCS projects to move forward without endangering underground sources of drinking water.

Thank you, Mr. Chairman and members of the committee for this opportunity to describe EPA's important work on carbon sequestration. I would be happy to answer any questions you may have.

Senator Domenici. Mr. Grumbles, I just have one question that came to my mind. I don't understand; why does it take 4 years to make a rule regarding a substance of like CO_2?

Mr. Grumbles. We have a lot of experience with CO_2 already under the UIC program under the class two category for enhanced oil and gas recovery. But this is a whole new approach, a whole new opportunity when it comes to long term storage. It's a complex issue and it requires public participation and discussion and also that 4 years will allow us to take into account lessons learned as the demonstration projects occur, whether they are DOE or other demonstration projects through the process.

Senator Domenici. All right. Thank you.

Mr. Allred.

STATEMENT OF STEPHEN ALLRED, ASSISTANT SECRETARY FOR LAND AND MINERALS MANAGEMENT, DEPARTMENT OF THE INTERIOR

Mr. Allred. Senator Domenici, members of the committee, it's a pleasure to be here and to represent the Department of the Interior. As you're aware, we deal with oil and gas, coal, geothermal and biomass as well as fish and wildlife issues on Federal lands and Federal resources. We're involved in a number of different strategies with regard to carbon sequestration, including the capture and storage through enhanced oil recovery of carbon dioxide and into geologic formations and also investigating the capture through air and terrestrial biomass, soils and trees of carbon dioxide on our lands.

One of the things that is probably not well known is the fact that currently carbon dioxide is a salable commodity under the mineral leasing laws of the United States. We currently collect revenues in the form of royalties from the sale of carbon dioxide produced in connection with oil and gas production on public lands. In 2007 the sale of carbon dioxide generated over 23 million dollars in royalty revenues in Colorado, New Mexico and Wyoming. I mention that because one of the things that I think as we look forward to carbon sequestration is that it may well be a valuable resource for the future that can be recovered as we need it.

Specifically I would like to talk a little bit more about the opportunity I think that we have to investigate and understand carbon dioxide sequestration because of the number of enhanced oil recovery projects that we have. Some of the largest new ones are in Senator Barrasso's area in Wyoming. I think we have an opportunity by adding perhaps another purpose to those projects to learn a lot from carbon sequestration that we can then apply to other areas. It's also interesting to note that there are estimates that we have the capacity, based on preliminary estimates of carbon dioxide production and volume in our oil reservoirs to sequester, if they were in the right places, 20 to 40 years of carbon dioxide production.

Public Law 110–140, enacted last year, gave the Department of the Interior a number of different responsibilities. We are to develop and are currently involved in the process of developing a methodology for and to conduct national assessments of geologic capacity for sequestration. We are in the process of preparing a national assessment of storage capacity of oil and gas and saline reservoirs through the USGS. We will conduct that in conjunction with the Department of Energy, with the input of the Environmental Protection Agency, and equally as important, the State geological surveys that we work with.

It is our intent to convene an independent panel of experts and stakeholders to provide a technical review of that effort. Upon completion of that review the methodology will be published and provided for public use. There are also a number of other things that are ongoing efforts that we'll also try to answer some of these questions, certainly to identify the criteria to determine candidate geologic sequestration sites in various different types of geologic settings.

We will also be involved in looking at a proposed regulatory framework for leasing decisions on public lands with regard to long term sequestration. We also were mandated to determine a means by which we can provide for public review and comment and to ensure that we protect the quality of natural and cultural resources in those areas. To suggest additional legislation that we feel might be important. Again I want to emphasize as we go forward with these we will consult with our Federal and State partners and the public.

We believe, in conclusion, that it is extremely important to No. 1, understand the effect and to determine the complex issues that we will have in sequestration and there are many interrelated components to that. Secondly, we believe that it is important to use the ongoing activities that are already in place to try to enhance the knowledge that we have. The Department of the Interior is looking forward to working with other agencies and working with Congress to answer these important questions. I'd be most happy to answer any questions. I have a couple of experts with me if I can't answer those today.

[The prepared statement of Mr. Allred follows:]

Prepared Statement of Stephen Allred, Assistant Secretary for Land and Minerals Management, Department of The Interior

Introduction

Mr. Chairman and Members of the committee, thank you for the opportunity to provide the Department of the Interior's views on S. 2323, the ''Carbon Capture and Storage Technology Act of 2007'' and S. 2144, the ''Carbon Dioxide Pipeline Study Act of 2007.'' As both bills vest the Secretary of Energy with primary authority and the Secretary of the Interior is identified as a cooperator, I will defer to the Department of Energy for specific views on this legislation. My testimony today will address the Department of the Interior's perspective on carbon capture and storage as it relates to future work of the Department's bureaus, specifically the U.S. Geological Survey (USGS) and the Bureau of Land Management (BLM).

The challenges of addressing carbon dioxide accumulation in the atmosphere are significant. Fossil fuel usage, a major source of carbon dioxide emissions to the atmosphere, will continue for the foreseeable future in both industrialized and developing nations. Therefore, a variety of strategies are being investigated to reduce emissions and remove carbon dioxide from the atmosphere. Such strategies include the facilitated sequestration of carbon for the capture and storage of carbon dioxide by injection into geologic formations as well as capture from the air to terrestrial biomass, including soils and trees.

Carbon injection techniques also have useful practical applications in processes known as enhanced oil recovery (EOR), which currently takes place on some public lands managed by the Bureau of Land Management. Carbon dioxide is a saleable commodity under the Mineral Leasing Act of 1920. The Bureau of Land Management currently collects revenues in the form of royalties derived from the sale of carbon dioxide produced in connection with oil and gas production on public lands. In 2007, for example, the sale of carbon dioxide generated over $23 million in royalty revenue in the states of Colorado, New Mexico, and Wyoming.

In addition to enhancing oil recovery, EOR's utilization of carbon injection may yield valuable data that will inform efforts to capture and sequester carbon dioxide effectively in geologic formations found on public lands. A critical issue for evaluation of storage capacity is the integrity and effectiveness of these formations for sealing carbon dioxide underground, thereby preventing its release into the atmosphere.

Geologic Storage of Carbon

The current atmospheric carbon dioxide concentration is approximately 380 parts per million volume and rising at a rate of approximately 2 parts per million volume annually, according to the most recent information from the Intergovernmental Panel on Climate Change (IPCC). The 2005 IPCC Special Report on Carbon Dioxide Capture and Storage concluded that in emissions reductions scenarios striving to stabilize global atmospheric carbon dioxide concentrations at targets ranging from 450 to 750 parts per million volume, the global storage capacity of geologic formations may be able to accommodate most of the captured carbon dioxide. How much of this carbon dioxide storage capacity would be economically feasible (assuming some price on carbon), however, is not known. Also, geologic storage capacity may vary widely on a regional and national scale. A more refined understanding of geologic storage capacity is needed to address these knowledge gaps.

Geological storage of carbon dioxide in porous and permeable rocks involves injection of carbon dioxide into a subsurface rock unit and displacement of the fluid or formation water that initially occupied the pore space. This principle operates in all types of potential geological storage formations such as oil and gas fields, deep saline water-bearing formations, or coal beds. Most of the potential carbon dioxide storage capacity in the U.S. is in deep saline formations.

Ongoing Efforts

H.R. 6, the Energy Independence and Security Act of 2007 (EISA), which the President signed into law last month, includes provisions on Carbon Capture and Storage that the Department is working to implement. The requirement in Section 7 of S. 2323 directing the Secretary of the Interior, acting through the Director of the U.S. Geological Survey (USGS), to develop a methodology for and conduct a national assessment of geological storage capacity for carbon dioxide is very similar to Section 711 of EISA and therefore we believe inclusion of this provision in new legislation is unnecessary.

The Department has developed an implementation plan for Section 711. In fiscal year 2008, the Department will begin development of a methodology that could be used to conduct assessments of carbon dioxide storage capacity in oil and gas reservoirs and saline formations nationally. The methodology development will be conducted in coordination with a number of organizations in order to maximize the use-fulness of the assessment for a variety of partners and stakeholders. These organizations include the Department of Energy, the Environmental Protection Agency, and State Geological Surveys. In particular, the Department will coordinate its work with Department of Energy's National Carbon Sequestration Database and Geographical Information System (NATCARB). The purpose of NatCarb is to assess the carbon sequestration potential in the U.S. and to develop a national Carbon Sequestration Geographic Information System (GIS) and Relational Database covering the entire U.S.

An independent panel, consisting of individuals with relevant expertise and representing a variety of stakeholder organizations, will be convened to provide a technical review of the methodology. Upon completion of the review, the methodology will be published and available for public use. The subsequent national assessment called for by EISA would need to compete among other administration priorities for funding.

In addition, Section 714 of the EISA directs the Department to develop a framework for geological sequestration on public land and report back to this committee, as well as the House committee on Natural Resources, by December 2008.

This effort, coordinated among several agencies within the Department, is anticipated to result in recommendations relating to:

- criteria for identifying candidate geological sequestration sites in several specific types of geological settings;
- a proposed regulatory framework for the leasing of public land or an interest in public land for the long-term geological sequestration of carbon dioxide;
- a procedure for ensuring any geological carbon sequestration activities on public land provide for public review and protect the quality of natural and cultural resources;
- if appropriate, additional legislation that may be required to ensure that public land management and leasing laws are adequate to accommodate the long-term geological sequestration of carbon dioxide; and
- if appropriate, additional legislation that may be required to clarify the appropriate framework for issuing rights-of-way for carbon dioxide pipelines on public land.

The report will also describe the status of Federal leasehold or Federal mineralestate liability issues related to the release of carbon dioxide stored underground in public land, including any relevant experience from enhanced oil recovery using carbon dioxide on public lands.

The report will, in addition, identify issues specific to the issuance of pipeline rights-of-way on public land and legal and regulatory issues specific to carbon dioxide sequestration on land in cases in which title to mineral resources is held by theUnited States, but title to the surface estate is not.

This effort will be undertaken in coordination with the Environmental Protection Agency, the Department of Energy, and other appropriate agencies.

Conclusion

It is clear that addressing the challenge of reducing atmospheric carbon dioxide and understanding the effect of global climate change is a complex issue with many interrelated components. The assessment activities called for in the recently passed Energy Independence and Security Act of 2007 should ultimately increase the information base upon which decision makers will rely as they deal with these issues, and the assessments called for in these bills would duplicate those already mandated. In addition to addressing the challenges presented by carbon dioxide, we should also recognize that this commodity presents certain opportunities for future knowledge and utilization. As a leasable commodity, our experience demonstrates that there is a demand and a value attributable to this resource. As we examine undeveloped oil and gas reservoirs, we should consider the potential benefits of accessible sequestered carbon dioxide. It is clear that the discussion on this subject will continue and the Department stands ready to assist Congress as it examines these challenges and opportunities. Thank you for the opportunity to present this testimony. I am pleased to answer questions you and other Members of the committee might have.

The Chairman [presiding]: Thank you very much. Mr. Slut, go right ahead.

STATEMENT OF JAMES SLUTZ, ACTING PRINCIPAL DEPUTY ASSISTANT SECRETARY, OFFICE OF FOSSIL ENERGY, DEPARTMENT OF ENERGY

Mr. Slutz. Ok. Mr. Chairman, members of the committee, it is my pleasure to appear before you today to discuss Senate bills 2323 and 2144. I currently serve as the Acting Principle Deputy Assistant Secretary for Fossil Energy.

Balancing the economic value of fossil fuels with the environmental concerns associated with fossil fuel use is a difficult challenge. Carbon capture and storage technologies provide a key strategy for reconciling energy and environmental concerns. DOE has assumed a leadership role in the development of carbon capture and storage technologies and in fact the United States, I would argue, is the global leader in this area. Through its carbon sequestration program DOE is developing the technologies through which geologic carbon sequestration could become an effective and economically viable option for reducing CO_2 emissions.

Since DOE first investigation into carbon sequestration began in 1997 with a budget of one million dollars, DOE has spent a total of 483 million through fiscal year 2008. Our fiscal year 2009 budget request of 148 million is a powerful sign of the continuing importance of this technology to our energy and environmental future. I might add that fiscal year 2009 we had a 30 million dollar increase over our 2008 request.

A recent report completed by the National Petroleum Council which was a comprehensive landmark study requested by the Secretary of Energy titled, ''Facing the Hard Truths About Energy''. The NPC's purpose of this chapter was to increase understanding about the scale and significance of the energy industries activities and to produce sound, balanced strategies to meet today's challenges and benefit future generations. In the section of the report dealing with carbon management, the NPC recommended and I quote, ''The United States must develop the legal and regulatory framework to enable carbon capture and sequestration and as policymakers consider options to reduce CO_2 emissions provide an effective global framework for carbon management.'' DOE is doing precisely that as a few examples more fully described in my submitted testimony illustrate.

DOE is working to increase the cost effectiveness of carbon capture technologies and to prove the viability of a long term geologic and terrestrial CO_2 storage. DOE's regional carbon sequestration partnerships are co-funding field tests for large scale CCS demonstrations. DOE is working closely with the Environmental Protection Agencies and the states through organizations such as the Interstate Oil and Gas Compact Commission to establish a standardized regulatory framework for CO_2 storage in deep geologic formations.

In December 2007, DOE participated in EPA's first workshop in preparation for a proposed rule for large scale injection of CO_2. Fortunately the United States has a large number of geologic formations amenable to CO_2 storage. In fact according to the 2006 carbon sequestration atlas of the United States and Canada, aggregate CO_2 sink capacity is estimated to hold several hundred years of total domestic U.S. emissions.

The U.S. Government, DOE and other agencies of the 50 States, several Canadian provinces, private industry, environmentalists and the scientists and the engineers have expanded great effort, invested heavily and made remarkable progress over the last decade in understanding and preparing for an energy and environmental future in which carbon sequestration technology will play an integral role. DOE believes regarding the specific bills being considered, DOE does have some specific positions. DOE believes that the research and development and demonstration projects prescribed in Sections 3, 4 and 6 of Senate Bill 2323 are duplicative of our R&D demonstrations underway in our existing program.

We also believe that Section 5ive which requires interagency task force duplicates the task force the EPA has underway. Section 6 of the proposed legislation would shift lead responsibility for part of the Department's R&D program from Fossil Energy to the Office of Science. DOE opposes this provision.

Senate Bill 2144 would require a feasibility study related to the construction and operation of pipelines and carbon dioxide sequestration facilities and for other purposes. DOE supports this legislation.

There are many questions and I think it's useful to consider the scale of CO_2 issues and CO_2 management. Building the infrastructure required to capture and store the CO_2 emitted by energy producing activities requires serious, long term commitment on the part of government and industry and the public. As an example, in the United States all the CO_2 from existing coal fired electric generation alone would total, if liquefied, would total 50 million barrels per day. That's two and a half times the volume of oil handled daily in the United States. Again I mention that just to scale the challenge and illustrates the urgency of moving forward with some of these issues.

Mr. Chairman, thank you for the opportunity to testify today. I'd be happy to answer any questions.

[The prepared statement of Mr. Slutz follows:]

Prepared Statement of James Slutz, Acting Principal Deputy Assistant Secretary, Office of Fossil Energy, Department of Energy

Mr. Chairman, members of the committee, it is a pleasure for me to appear before you today to discuss Senate bills 2323 and 2144.

I intend, first, to survey The Department of Energy's (DOE) overall Carbon Sequestration Research and Development program, our goals and our progress to date. I will then describe DOE's collaboration with the Environmental Protection Agency (EPA) in carbon capture and storage.

Complete knowledge of all these efforts already underway should be of interest as the Senate bills under consideration go forward.

Carbon Sequestration and Fossil Fuels

The availability of affordable energy is a bedrock component of economic growth. The use of fossil fuels, however, can result in the release of emissions with impacts on the environment. Of growing significance are emissions of carbon dioxide (CO_2) which contribute to global climate change.

Balancing the economic value of fossil fuels with the environmental concerns associated with fossil fuel use is a difficult challenge. Carbon capture and storage technologies provide a key strategy for reconciling energy and environmental concerns. Geologic sequestration—the capture, transportation to an injection site, and long-term storage in a variety of suitable geologic formations—is one of the pathways that DOE is pursuing to allow the continued use of fossil fuels while reducing CO_2 emissions.

DOE has assumed a leadership role in the development of carbon capture and storage technologies. Through its Carbon Sequestration Program—managed within DOE's Office of Fossil Energy and implemented by the National Energy Technology Laboratory (NETL)—DOE is developing technologies through which geologic carbon sequestration could potentially become an effective and economically viable option for reducing CO2emissions. The Carbon Sequestration Program works in concert with other programs within the Office of Fossil Energy that are developing the complementary technologies that are integral to coal-fueled power generation with carbon capture: Advanced Integrated Gasification Combined Cycle, Advanced Turbines, Fuels, Fuel Cells, and Advanced Research. Successful research and development could enable carbon control technologies to overcome various technical and economic barriers in order to produce cost-effective CO_2 capture and enable wide-spread deployment of these technologies.

DOE's Carbon Sequestration Program

Since DOE's first investigation into carbon sequestration began in 1997 with a budget of $1 million, DOE has spent approximately $483 million through Fiscal Year 2008 (twelve year cumulative total) on further research and development, a powerful sign of the importance of this technology to our energy and environmental future.

The Carbon Sequestration Program, with a Fiscal Year 2008 budget of $119 million, encompasses two main elements of technology development for geologic sequestration: Core R&D and Validation and Deployment. The Core R&D element addresses several focus areas for laboratory technology development that can then be validated and deployed in the field. Lessons learned from the field tests are fed back to the Core R&D element to guide future research and development. Through its Integrated Gasification Combined Cycle, Fuels, Sequestration, and Advanced Research programs, DOE is investigating a wide variety of separation techniques, including gas phase separation and adsorption, as well as hybrid processes, such as ad sorption/ membrane systems. Current efforts cover not only improvements to state-of-the-art technologies but also the development of several revolutionary concepts, such as metal organic frameworks, ionic liquids, and enzyme based systems. The ultimate goal is to drive down the energy penalty associated with capture so that geologic sequestration can be done with only a moderate increase in the cost of electricity.

Regional Carbon Sequestration Partnerships

One of the key questions regarding geologic sequestration is the ability to store CO_2 in underground formations with long-term stability (permanence); this requires monitoring and verification of the fate of the CO_2, to ensure that the science is sound and ultimately gains public acceptance. DOE's NETL, with the Regional Carbon Sequestration Partnerships (RCSPs) are developing and validating technology, and national infrastructure needed to implement geologic sequestration in different regions of the Nation.

The RCSPs are evaluating numerous geologic sequestration approaches in order to determine those best suited for specific regions of the country. They are also helping develop a framework to validate and deploy the most promising technologies for geologic sequestration.

A Three-Phase Approach

NETL's three-phased approach began with a Characterization Phase in 2003 that focused on characterizing regional opportunities for carbon capture and storage, and identifying regional CO_2 sources and storage formations. The Characterization Phase was completed in 2005 and led into the current Validation Phase, which focuses on field tests to validate the efficacy of geologic sequestration technologies in a variety of storage sites throughout the U.S. Using the extensive data and information gathered during the Characterization Phase, NETL identified the most promising opportunities for carbon storage in their regions and commenced geologic field tests. In addition, NETL is verifying regional geologic sequestration capacities initiated in the first phase, satisfying project permitting requirements, and conducting public outreach and education activities.

The third phase, or Deployment Phase, for large-volume testing is intended to demonstrate the feasibility of CO_2 capture, transportation, injection, and storage at a scale comparable to future commercial deployments. DOE has in recent months awarded funds to initiate five large-volume demonstration projects. Depending on the results of a scientific needs assessment being conducted in FY 2008 and the ability of additional project proposals to meet those needs, additional large-scale projects may be initiated. In October, 2007, DOE announced awards totaling $318 million for two projects with the Plains Carbon Dioxide Reduction Partnership, and one project each with the Southeast Regional Carbon Sequestration Partnership and Southwest Regional Partnership for Carbon Sequestration. In December, DOE announced a $66.7 million award for a project with the Midwest Geological Sequestration Consortium.

The geologic structures to be tested during these large-volume storage tests will serve as potential candidate sites for the future deployment of technologies demonstrated in FutureGen and the Clean Coal Power Initiative, which plans to complete a solicitation for carbon capture technologies at commercial scale in 2008.

The NETL, with the RCSPs and the National Carbon Sequestration Database and Geographical Information System (NATCARB), has created a methodology to determine the capacity for CO_2 storage in the United States and Canada and an Atlas from data generated by the RCSPs and other databases, including the United States Geological Survey's (USGS) National Coal Resources Data System, USGS National Water Information System Database, and EROS Database. Based on data displayed in the *2006 Carbon Sequestration Atlas of the United States and Canada,* the aggregate CO_2 sink capacity—including saline formations, unmixable coal seams, and oil and natural gas formations—is estimated to hold several hundred years of total domestic U.S. emissions.

Moving Toward Commercial Deployment

Carbon capture and storage can play an important role in mitigating carbon dioxide emissions under potential future stabilization scenarios. The United States has a large capacity of geologic formations amenable to CO_2 storage. DOE's Carbon Sequestration Program will continue to help move geologic sequestration technology toward readiness for commercial deployment.

EPA's Role in the Deployment of Carbon Capture and Storage Technology

Complementing DOE's carbon capture and R&D research program is the EPA program for ensuring that underground injection of CO_2 is conducted in a manner that is protective of underground sources of drinking water (USDWs) in accordance with section 1421(d)(2) of the Safe Drinking Water Act (SDWA). EPA is initiating workto develop proposed regulations to ensure consistency in permitting commercial scale geologic sequestration projects. It plans to propose regulations in the summer of 2008. EPA is also responsible for reviewing and commenting on environmental impacts statements under the National Environmental Policy Act (NEPA).

As DOE moves forward with its R&D program and geological storage projects, EPA is focused on: evaluating risks to human health and the environment; providing guidance on permitting CO_2 injection wells for pilot-scale projects; identifying technical and regulatory issues associated with field tests and commercial projects; and developing an appropriate management framework for permitting.

DOE-sponsored and industry-sponsored research will help develop data and tools to address these issues. It is anticipated that EPA will aggregate and analyze the information generated from those efforts and initiate new research where there are gaps.

DOE has also sponsored a five-year, two-phase study by the Interstate Oil and Gas Compact Commission (IOGCC), which is reported on in the publication a Model CO_2 Storage Statute and Model Rules and Regulations. The report provides industry perspective on development of regulations governing the storage of CO_2 in geologic media and an explanation of those regulatory components. EPA will consider theseand other viewpoints in its regulatory development process.

Program Coordination

EPA coordinated with DOE in the preparation of its research plan, and is working closely with DOE, state regulators and other stakeholders on all geological storage activities so as to leverage resources, clarify key questions and data gaps, and ensure that work is complementary and not duplicative.

EPA and DOE, for example, hold quarterly coordination meetings (at both the staff and managerial level) to share progress and discuss key issues.

EPA, in coordination with DOE, organized a series of technical workshops in 2007 to help define future research needs. The workshops were focused on technical issues that need to be addressed in order to design, operate, and permit CO_2 injection wells. Attendees included EPA and state regulators, DOE project managers, and DOE-funded researchers.

In addition, EPA has and will continue to be involved in major DOE/NETL activities such as the National Conferences on Carbon Sequestration and the Regional Partnership Annual Review Meetings.

S. 2323: Carbon Capture and Storage Technology Act of 2007

The U.S. Government, DOE and other agencies, the 50 states, several Canadian provinces, private industry, environmentalists, and scientists and engineers have expended great efforts, invested heavily and made remarkable progress over the last decade in understanding and preparing for an energy and environmental future in which carbon sequestration technology will play an integral role.

The Administration strongly supports research and development of carbon capture and storage technology as a solution to reduce carbon dioxide emissions and address global climate change. The Administration is currently performing the research and development needed to successfully develop this technology. DOE has numerous initiatives looking at decreasing the cost of carbon dioxide capture and proving the permanence of carbon dioxide storage in geologic formations and has success with its current structure. DOE believes that the research, development and demonstration projects prescribed in Sections 3, 4 and 6 of Senate Bill 2323 are generally duplicative of the R&D and demonstrations underway in our existing program. DOE is currently evaluating some of details of this bill within the context of its existing program, such as the use of competitive grants to fund commercial demonstration of carbon dioxide sequestration and the number of projects needed.

Section 5 of this bill would require an interagency task force to develop regulations for the capture and storage of carbon dioxide. This task force was officially established last year, and is chaired by EPA, with considerable support from DOE. Therefore, we believe that this section of the bill is also redundant.

For the past 10 years, DOE's Sequestration Program within the Office of Fossil Energy has funded research in areas of carbon dioxide capture, storage, monitoring, mitigation, and verification (MMV), breakthrough concepts, and infrastructure development through its Regional Partnership Initiative. NETL is researching the most suitable technologies, informing regulatory development, and evaluating infrastructure needs for carbon capture, storage, and sequestration in different areas of the country. The RCSPs are conducting much of these efforts, and include 41 states and over 350 distinct organizations working together for the most cost-effective solutions. Additionally, the Clean Coal Power Initiative and FutureGen are providingthe demonstration platform for testing larger carbon dioxide capture methods at power plants. These activities are currently providing the plan forward and should continue along their current path to produce the best results at the earliest time so that this technology can be an important option to cost-effectively reduce greenhouse gas emissions.

S. 2144: Carbon Dioxide Pipeline Study Act of 2007

This bill would require the Secretary of Energy, in consultation with the Federal Energy Regulatory Commission, the Secretary of Transportation, the Administrator of the EPA, and the Secretary of the Interior to conduct a feasibility study relating to the construction and operation of pipelines and carbon dioxide sequestration facilities, and for other purposes. It also requires that the Secretary provide this chapter to Congress no later than 180 days after the enactment of this bill. DOE supports this legislation and notes that, although it is the study lead, it will work closely with the other agencies in conducting this study, and in particular with DOT's Pipeline and Hazardous Materials Safety Administration (PHMSA), which will have a leading role in evaluating plans for construction and operation of pipelines for carbon dioxide. Mr. Chairman, and members of the committee, this completes my prepared statement. I would be happy to answer any questions you may have at this time.

The Chairman. Thank you all very much. There are a lot of questions, obviously. One of the questions that occurs to me is what we are talking about when we talk about permanent CO_2 storage. I know Senator Kerry referred to permanent CO_2 storage.

Mr. Grumbles, I was not here during your testimony. I had to step out, but is this an issue? Is EPA planning to clearly define the required length of storage time for CO_2 in the regulatory work that you're working on these days?

Mr. Grumbles. Senator, I know we will certainly address that issue. That is one of the key areas, long term liability, financial responsibility and duration of requirements. So we do intend to seek comment on questions such as those about the duration and also on the responsibility, financial responsibility requirements.

Currently much of the UIC program, Underground Injection Control Program, has in place for other types of injections of fluids, financial responsibility requirements that last typically 30 years. So one of the questions that is very much going to be at the forefront of our debate and discussion is duration. How long of a period to ensure that there's monitoring and responsibility, financial responsibility.

The Chairman. Let me ask you, Chairman Kellier, what road do you see FERC as appropriately performing here? Is FERC the appropriate entity to regulate transportation rates? Is that something that you think is clearly FERC's responsibility?

Mr. Kelliher. I think we could regulate the rate. There are three really existing models on how to regulate pipeline transportation either of energy resources or CO_2. In two of the three existing models FERC does set the transportation. In the other one the rate is—can be set by the Surface Transportation Board upon complaint. Otherwise it's set through a contract between parties. But FERC certainly has set the rate for oil and gas pipelines for decades.

The Chairman. Let me ask you, Mr. Allred, if you could just elaborate a little bit on what responsibility you would see in your agency for the site selection on Federal lands to the extent that someone were to determine that they wanted to pursue a carbon storage sequestration project on Federal lands. How far are you from having in place a regulatory framework that would tell them how to proceed?

Mr. Allred. Mr. Chairman, members of the committee, I think that at this point in time we would have the basic rules that we could use in a sequestration project. There will be lots of questions as we go forward. One of the things that we have thought about are some demonstration projects that would, associated with enhanced oil recovery, help us answer some of those questions.

But I think the leasing rules and the Mineral Leasing Act and the laws that we have with regard to the management of Federal lands provide us the basics that we would need. That does not mean that as we go forward there won't need to be enhanced and changed.

The Chairman. All right. Let me go ahead and call on the next Senator on this.

Senator Barrasso.

Senator Barrasso. Thank you very much, Mr. Chairman, and I want to first commend Senators Coleman and Senator Kerry for bringing this legislation before us today. The carbon capture, transmission and sequestration are enormously important to the people of Wyoming. Mr. Allred made some comments about that.

Wyoming produces about 38 percent of our Nation's coal. This coal provides a substantial portion of America with affordable electricity. Wyoming holds promising geologic formations for coal sequestration. Wyoming has experience in safely moving carbon dioxide and effectively using it for our enhanced oil recovery. Since emissions are not always located near appropriate geologic voids, carbon dioxide may need to be transported through States like Wyoming.

So I think we need some additional research to inform us on these issues. Congress and State Legislatures must fully explore any gaps in the existing legal and regulatory frameworks. The Wyoming Legislature is doing that when it meets next week in Cheyenne for this session.

I concur with Senator Coleman and Kerry on the urgency of this problem—and certainly in light of the current climate change debate and the potential of Congress imposing a price on carbon in one form or another. I am hearing a lot about this around the State of Wyoming. I'm hearing it from workers, from consumers, and also from industry.

Many of my constituents are increasingly concerned that the Federal Government will impose a cap on carbon before we've developed the appropriate legal and regulatory framework to address carbon dioxide. They're concerned that Congress is going to act on those issues before we act on the technology. So I applaud what Senators Coleman and Kerry are trying to do.

If anything I'm going to be pushing to do more research and more analysis, find more certainty for markets and to delve deeper into the important areas such as liability of CO_2 as it's transported and stored and some of the things that we've heard from the Honorable Edwards and to accomplish all of this in a shorter period of time. I'm distressed when I hear it's going to take till the year 2011 before regulations can be written. I want to thank the Chairman for calling this meeting.

I do have a question or two for Mr. Slutz. It really has to do a little bit with something along this line but a little different and that's the Future Gen issue from the Department of Energy. We talk about public/private partnerships, and when private entities put up money and make decisions, business decisions, that they want to have confidence that they're going to have a good partner in the U.S. Government. With this change for Future Gen, I think that confidence is eroding.

My question is how can we make sure that commitments by the Federal Government are really commitments that the private enterprises can depend upon and rely on?

Mr. Slutz. Thank you, Senator. You know and the, I should say the Future Gen decision that was announced yesterday was a very challenging and difficult decision. But part of that commitment is a partnership and you have to have both—there's the Federal Government and the industry partners. Both of those have to have agreements that they can work through and when a project doubles in cost it's a time to sit down and rework those agreements. In the details of that it was decided that the current agreement was not in the best interest of the American people and there were—the requirements that would have been in place to go forward would have put that project at risk.

So this issue is very, very important, too important not to allow that not to be successful. The decision was made by the Department to—with a new direction. Not a lack of commitment to Future Gen or a lack of commitment to furthering research and furthering demonstrating the technology of carbon capture and storage on a full commercial scale project, but doing it a way, I think, that the market could really act and it would be very market based commercial type plans that would be where the success would be very likely. That was the foundation for that decision is to make sure we have a successful outcome.

Senator Barrasso. Mr. Chairman, as a member of both the Energy committee as well as the Environment and Public Works committee we're looking at using all the sources of energy, but doing it in an appropriate way for our environment. We don't want to put the cart before the horse too far. We want to make sure that we're working in unison to try to get things going on developmentally and educationally in a way that would help folks and help our environment, but also make sure that we have all the energy that we need. Thank you, Mr. Chairman.

The Chairman. Thank you very much.

Senator Tester.

Senator Tester. Thank you, Mr. Chairman. I want to thank Senator Barrasso for his comments. I want to dovetail on to them initially. Mr. Slutz, what's the level of urgency in the DOE as far as carbon capture and storage?

Mr. Slutz. It's the very critical to much of our research programs. A majority of clean coal is about how do we increase efficiency, reduce emissions or actually have carbon capture and storage. I think one of the urgencies that's communicating if you look at part of this decision on Future Gen and to communicate the importance that the Administration places on our clean coal research program.

We released our budget numbers early this year and showed a very significant increase in coal research which shows how we want to move this faster.

Senator Tester. Coal research based on carbon capture?

Mr. Slutz. Based on carbon capture and storage.

Senator Tester. How close are we to having a large scale carbon capture technology available?

Mr. Slutz. The technology exists. The issue is the cost of the technology and the costs that it would——

Senator Tester. How close are we to having to having affordable large scale carbon capture technology?

Mr. Slutz. First let me, the technology exists. One of the advantages of this Future Gen approach is to actually get a commercial power plant in operation. We're looking at a date of 2015 that it could be operational on this because we don't know all of the costs. We only now can speculate and estimate the costs, but having actual real data on commercial scale, commercial operating power plants will be hugely beneficial in making those advancements in the future.

Senator Tester. Ok. I just tell you I echo Senator Barrasso's comments. We have to have the technology. That technology is critically important. Everybody blames the price of food on corn ethanol. There's also a worldwide drought. That probably has more to do than any kind of renewable energy policy we put out of here.

I attribute a lot of that to worldwide global warming. I think that if we lead, and we lead in a way that's there's urgency, we can have our eggs and eat it too, as Senator Coleman said. That is that we can develop technology we can sell to other countries and we can clean it up on a worldwide basis. But there has to be a level of urgency. Because if there's not a level of urgency, we'll develop the technology and it will be too late.

I've got a question for Mr. Grumbles from the EPA. You talked about under any proposed regulation, and I'm reading from your written statement, currently under development that geological site characterizations to ensure that wells are sited in suitable areas to limit potential for migration and injected formation fluids into underground drinking water. Now I know you're with the drinking water area of the EPA, but what about leeching into the atmosphere? Isn't that also considered insuitable sites?

Mr. Grumbles. It's a very important issue that is part of our discussions with the Air Office. It's part of our discussions with all the stakeholders so far and all the national workshops and research forums we've had. We want to look at the environmental risks but from the standpoint of the Safe Drinking Water Act.

Senator Tester. Ok.

Mr. Grumbles. We will be focusing on protection of underground sources of drinking.

Senator Tester. I appreciate that. But so what you're saying here only as it applies to drinking water, but somebody's got to be looking at leeching of the atmosphere because that's another significant problem.

Mr. Grumbles. Monitoring of these sites it's very important to have post closure monitoring of sites as it is with existing UIC programs.

Senator Tester. Ok.

Mr. Grumbles. Monitoring can get into releases, surface releases as well.

Senator Tester. Good. Mr. Allred, I have a question that kind of dovetails on the chairman's question. It deals with storage and with the plume that will be created with CO_2 storage. I don't care where you put it. That plume may affect not only Federal lands but also private lands and private lands where there's Federal minerals underneath them.

What do you recommend, or do you have a recommendation to deal with the liability as it applies to the plume and potential exposure there. Keeping in mind that, we put it under the ground, somebody may decide to drill a hole and let it out, that might happen on private lands. So how do you deal with that issue?

Mr. Allred. Mr. Chairman, Senator Tester, I—as I look at the Federal land question and I look at most of my background as you probably know is not in government. I look at how do you implement projects or how do you cause things to occur. To me one of the most important things that we need to look at is, first, to keep regulation as simple as it can be because if it is complicated, I think that that will deter people and organizations and financial resources from being applied to this.

Second, there's a big learning curve on this. The more that we can use existing legal frameworks to assure that we don't have unintended consequences or that we provide the ownership and responsibility for these the better off we'll be. I think that our oil and gas laws, both the Federal Government's and the State's, perhaps provide some of that mechanism. Particularly on Federal lands we think that through our leasing requirements and steps that we take to assure that oil and gas is properly handled and that the United States receives its proper royalty from that.

If you use the reverse of that, we think that those same processes or same legal applications can work on Federal grounds. I think one of the things you really have to consider is who owns that carbon dioxide because if you decide that then you probably have gone a long way to decide about liability and responsibility. If they are on Federal lands or Federal resources we're either going to have an agreement with the lessee to store their carbon dioxide or there will be decisions that that will become a Federal resource where there's Federal ownership and before that agreement is ever entered into there will be some assurances as to how safe it is to have it where it is.

Senator Tester. I've run out of time, but the other question is what happens if it starts on Federal land and moves into private land? I don't want you to answer that because there are other people that have questions but that is a concern.

The Chairman. Senator Corker.

Senator Corker. Actually I think that is a great question. This is an area that I guess I'm somewhat skeptical about just because, for a long, long time we've been pumping carbon in the air and we figure out here recently that's a problem. I don't know if we've done enough work to see what kind of problems will exist underground pumping tremendous amounts of carbon in.

But I'd love for you to answer Senator Tester's question because it seems to me that we do get into a lot of mineral rights issues, and storage rights issues. Storage can end up taking place under somebody's land and they might want to drill for something else. I mean it does seem to me there are a lot of complications that exist. So please answer Senator Tester's question.

Mr. Allred. Mr. Chairman, Senator Corker, to the best that I can, I will. But one of the advantages we have with depleted oil or even operating oil reservoirs is there's a tremendous amount of information about that reservoir with regard to its extent and characteristics. That's a source of data that we don't have perhaps in places that we might seek to sequester where we have not had that oil or gas experience.

Not in all cases will we be 100 percent sure. That's going to be a, I'm mean, there's got to be a concern about that because we may not. But remember why the oil and gas is there. It was captured because there was some structure that kept it there. What will be a bigger question probably is what have we done to make it not suitable because obviously we've drilled holes. That is an issue that can be dealt with fairly easily although it might be expensive.

The second question is when you enhance oil recovery, you do a thing called fracking and you fracture some of the area in order to have oil flow into a well more easily. That will be a real question as to how that fracking has affected those reservoirs and their integrity. That is something that we'll have to understand.

Senator Corker. So is it envisioned that if a property, private property owner has an area underneath them where oil has been recovered and now there's a cavern there or someplace just to store carbon that that person would actually be paid for carbon to be stored under their land?

Mr. Allred. Mr. Chairman, Senator, if you were to follow the oil and gas laws and that's a big if, I think.

Senator Corker. Right.

Mr. Allred. That has to be decided on non-Federal lands. I would assume that there would have to be a lease like an oil lease so you could use that. But I think that's one of the big questions about how do you regulate. I think the question on Federal land may be more clear than on private lands.

Senator Corker. Yes, I would guess so. I actually think that as we get into this, while I agree with others there should be a tremendous sense of urgency, I think there are so many complications that exist around this that I can see how it might take several years to work out this whole process of rulemaking even though we'd like to see it happen more quickly.

In the area of drinking water itself it seems to me that water is becoming more and more of an issue even in a State like Tennessee. Talk to us about some of the hazards that exist, that you can envision existing storing carbon adjacent to water supplies and those types of things.

Mr. Grumbles. Senator we have extensive experience as was already mentioned with the Class two UIC program under the Safe Drinking Water Act injecting carbon dioxide for enhanced oil and gas recovery. But one of the key areas of focus for us, as we work to issue a regulation ensuring there are safeguards for the long term storage and sequestration of carbon dioxide, is to reduce the likelihood of migration of that CO_2 into underground sources of drinking water. That it can lead to different types of pollution problems.

I would say one of the areas we're looking at as potential risks to aquifers, potential sources of drinking water, because we recognize that in some areas—data we have indicates that by 2013 at least 36 States in this country will experience some form of water shortage. That's not just in drought stricken areas. It's a combination of growth and population and development or drought.

So for us as we're going through this analysis, of CO_2 storage, we want to make sure that it doesn't migrate, that it stays in place. The experience to date is that it is a very promising technology. That it does stay in place for very long periods of time if the geologic siting is done properly, the well construction is done properly and it's monitored. So it is very promising in not producing or posing a significant risk to underground sources of drinking water.

Senator Corker. Mr. Chairman, thank you for the time. At some point I hope we'll gain an understanding as to the practicality of this. I think that a lot of times we move in a direction and it sounds utopic to move there and there's a lot of political momentum.

I have to tell you that this still, to me, and I'm not an oil drilling State, seems like a fairly impractical thing to do on a mass scale. I can see in geographic locations that might be good, but thank you so much for this hearing.

The Chairman. Senator Dorgan.

Senator Dorgan. Mr. Chairman, thank you very much. In North Dakota we have the only coal gasification plant that was ever completed and it exists now as a technological marvel in terms of its production. It produces synthetic natural gas from lignite coal. It also is, I think, the world's largest demonstration of CO_2 capture and it captures 50 percent of the CO_2 from the coal gasification plant. It puts it in a pipe and pipes it to Canada to the oil wells in Alberta and they use it for enhanced oil recovery.

So there is a capture process with respect to coal gasification. I suspect if we're taking a look at what's happening in the oil wells up in Canada we can get a sense of some leeching issues and other issues about is this sequestration for the long term and so on. But I want to ask this question. This is such an important area. Fifty percent of our electricity comes from coal. We're not going to have a future without using coal. I mean our future's going to include the use of coal. The question is how do we use coal.

We're going to be able to use coal if we can effectively unlock the mysteries of capturing carbon and sequestering carbon. How do we do that? I'm chairing the appropriations side on the subcommittee on appropriations that funds a lot of this. Here's the way it looks to me.

We've got carbon sequestration R&D projects. We've put 120 million into that. We have regional organization. So we got 120 million dollars in our area. It's called PECORE. But 120 million dollars was for carbon sequestration R&D. 70 million dollars for the clean coal power projects. 75 million dollars was put in this year for Future Gen, now I saw your announcement yesterday about that, six billion dollars for loan guarantees for coal projects that will demonstrate these technologies.

So we have all of these things happening. You know, it reminds me kind of a circus with a bunch of rings. The question is who brings it all to the center ring? Who brings it to center stage to decide how all of this works together because we're in a big hurry? The fact is we need to get to center stage soon with technologies we know will work. So, Mr. Slutz, who down there at the Energy Department is bringing these five or six projects or areas of funding together to accomplish what we want to accomplish?

Mr. Slutz. It is, Senator. We do manage these programs as an integrated approach even though it may not look like it all the time, but we do. In fact I will use an example in answering some of the questions about what's this going to do in the subsurface. That's why we have those—that carbon sequestration program with the seven partnerships and we have four large scale projects under-way. Three more will be finalized and ready to announce later this spring.

Those—the four that are announced are each there in the site characterization doing those detailed geologic assessments. They will when they, a little after that assessment is done and assuming the assessment all proves out, they are scheduled—will inject at least a million metric tons a year of CO_2. They'll be monitored extensively. It's to get that information, that detailed information, that then working with our partners at EPA and the other agencies can use to develop what's the best rules.

One side on that this CO_2 is not in some big cavern. It's in the pore space of rock. I think that's an important piece and I'm sorry to end——

Senator Dorgan. You know, last year I called down because what was happening inside the department is the Department of Science was over here and others were over here and the money wasn't being released to the partnerships. So I don't know that both hands were communicating so well. They finally got the money but there is an urgency about this.

Someone just gave me this core sample. This is sandstone and this is where you would invest CO_2. The question is does it stay there? What are the conditions under which it stays there?

You know, the science is very sophisticated, very important and my great concern is we're moving very quickly on this issue of climate change and our understanding that we have to take immediately no regret steps to deal with it and perhaps more aggressive steps. But that doesn't mean that this country's going to be able to have the kind of energy supply that it wants and needs without using our most abundant resource and that is coal. So, when I mentioned five or six programs I don't know that you have this all laced up real tight down there. I hope so.

Because just for example the last six billion dollars we put in which would be loan guarantees. I don't know how you intend to use those. I don't know the announcement you made yesterday. I don't know what that means in terms of the several larger projects rather than one in Future Gen. I don't know how that relates to the regional partnerships.

So, I hope you can continue to give us as much information as possible about how all these things come together and lace up something that gets us to a conclusion. Because we can—you know one of the things about the government is it just studies things forever. That's really interesting but not very effective. At some point you have to have coordinated studies that get the results you need in order to move forward and achieve the goals you have.

Mr. Slutz. Let me answer that as I think it requires a little more of an answer that we can provide outside in subsequent to the hearing. But it is coordinated even as we work through our budget process. In looking we've done climate overlays that look at things like where can we spend the money to get different benefits at different times from the various programs within DOE. Not just fossil, but energy efficiency and all those are layered together to see how would we address, how would the overall DOE research budget address climate change?

Then as you then work within the programs, for instance the fossil energy, we have the core coal R&D that's going to move these advance technologies. The CCPI program that gets those technology pieces out into a demo environment. The Future Gen project that is a full scale powered plant with CCS and then you mentioned you get these proved up in demo. You still have to get them deployed into the marketplace and that's where loan guarantees come into play.

So it is a program that works together. I would be happy to follow up with a little more detail on how that meshes up a little bit.

[The information follows:]

Attached is a graphic** that outlines the activities of the Office of Fossil Energy's Carbon Sequestration Program and the relationship of those activities to basic research carried out under DOE's Office of Science. Carbon Sequestration an Storage is one of six areas highlighted in the FY 2009 DOE budget request for enhanced coordination between basic and applied research and development. Coordination of activities between the DOE programs is carried out through program manager-level working groups.

Senator Dorgan. My time has expired. I want to thank all the witnesses today. I have to go to another hearing as well, but I appreciate very much your testimony.

The Chairman. Thank you.

Senator Murkowski.

Senator Murkowski. Thank you, Mr. Chairman. Thank you for the testimony here this afternoon. I want to follow on a little bit to Senator Dorgan's comments and those made by Senator Barrasso about the Future Gen project and the decision coming out of DOE and more specifically to the signal that is sent.

Again you had a pretty ambitious private/public partnership there. A lot of time, a lot of money goes into it and then I understand what you were saying, Mr. Slutz about having to re-evaluate and do the cost. But in terms of signals sent, it is not a very encouraging signal sent from the Department of Energy about one of these proposals that we're looking to firmly establish that the technology has worked and we're making it happen right here.

Ms. Edwards, I want to ask you a question about just the logistics of how these CO_2 pipelines would work. If we were to go to full CCS for all powers I would imagine we're going to have to expand this pipeline network, not only the numbers of pipelines, but the lengths of the pipelines that we're talking about. The question is, as you are pumping the CO_2 further distance and in larger volumes, does this cause any problems? You kind of spoke to the safety aspect of the process, but recognizing that it's going to be more, higher volumes going farther distances, does that do anything to us?

Ms. Edwards. Again these are the sorts of risks that we regularly manage in transportation. So I would think that the—we have a regulatory framework in place that's ready for CO_2 pipelines however they were configured. They may be short. They may be longer and with more capacity. So I would say that the——

Senator Murkowski. But it does cause the pipe to age more quickly?

Ms. Edwards. You know again it would and that would be part of the requirements or if it did it would be part of the requirements. You know the operator understands the risks and does the monitoring. Of course it's very consequence phased depending on where the segments run and what is the exposure.

Senator Murkowski. So do we not know yet?

Ms. Edwards. This is material that does not have the same types of, you know, certainly environmental risks or even risks of life and property as other materials that are moved by pipeline. So we, yes, you know those questions—the technology is mature for pipelines. So I would say that the core significant issues having to do with pipeline transportation are not technical but more economic in terms of siting and investment.

Senator Murkowski. Let me ask you this, Commissioner Kelliher, about the siting you mentioned in your opinion that the State siting seem to be working relatively well if you have an increase again in this pipeline network. With the siting issues, does it become more complicated with an increased capacity there in terms of the siting? Is it something that at some point you might say that the Federal preemption is the way to go? Is this something to be evaluated later?

Mr. Kelliher. I think looking at what Congress did on gas pipe-line siting. It started off with state siting and at some point it failed. In the views of Congress they concluded that state siting had failed. But it was after the failure was demonstrated. Then Congress came in and changed the law. Exclusive and preemptive siting was the rule.

The State siting has worked for CO_2 pipelines up to this point. But the network is much smaller than the oil and gas pipeline networks. The CO_2 pipeline network is about 3,900 miles. The natural gas pipeline network is about 300,000 miles and the oil pipeline network is about 200,000 miles. Last year actually was a very big year for gas pipeline additions and we added about 2,700 miles last year alone on gas pipelines.

So it really relates to if this is the path the country goes down how big of a CO_2 pipeline network are we going to need and how quickly are we going to need it? There are varying estimates on how big of a network we might need.

Senator Murkowski. So at some point in time you may have to evaluate this preemption issue. I would certainly think there would be state consultation if that is the route you would go. I want to ask you a question a little bit off topic, but since you're here, Commissioner Kelliher I'd like to ask you about this MOU that I understand this is regarding ocean energy projects.

You know I've been following this and trying to see some progress with this. There's been some competing Federal jurisdiction out there, FERC, saying anything within three miles MMS is looking at those projects located on the outer continental shelf. Can you give me a status very, very quickly as to what is happening with that MOU?

Mr. Kelliher. The quick status is that in my view it is final. In my view we're prepared to sign, but the MOU cannot be effective with only one agency signature.

Senator Murkowski. Can we expect the other agency signatures shortly? What is the status on that?

Mr. Kelliher. I would have to defer to Secretary Allred on that.

Senator Murkowski. Secretary, can you give me an update on where you are on your end?

Mr. Allred. Senator Murkowski, as you are aware we had negotiated an agreement. I think it's one that was generally acceptable to both of us. We were asked by this committee not to proceed with that in light of and until there were certain decisions that I assume the committee chose to make with regard to the energy legislation.

One of the things that's happened in the intervening period of time is that our regulations for alternative energy offshore are now about ready to be released. While I don't anticipate that that would make a significant change in how we and FERC deal together it's probably premature for us to do that in light of the fact that these regulations will go out for public comment. If the regulations themselves don't deal, and there are some that do deal with some of the issues in the memorandum of understanding, it would be my intent that we would modify that agreement so that it will be specific to the items that we might have yet uncovered in those rules and regulations.

Senator Murkowski. You're not giving me any indication in terms of timing on this then.

Mr. Allred. I would anticipate that the rules and regulations will be out for draft review within the next 2 months.

Senator Murkowski. We certainly—an awful lot of people that are hoping that this gets resolved very quickly and we're very optimistic when this MOU was announced. So, I'd like to think that we're going to see that sooner than later. My time is up, but perhaps I can have a little follow up after this with you if it's possible to do that. Thank you, Mr. Chairman.

Mr. Kelliher. Just 30 seconds or so, Mr. Chairman, thank you. I just want to emphasize how important it is that we clarify the respective roles of the two agencies. Actually I don't see that there's conflict between the MMS role and the FERC role. I actually think they're complementary. It's just that the two agencies actually have never worked together on projects. We haven't seen ocean projects in the outer continental shelf before that are FERC jurisdictional.

So from the developers point of view there's tremendous uncertainty. I think perpetuating that uncertainty is forestalling exactly what we need now on ocean energy demonstration projects. We need to demonstrate these technologies and the uncertainty means we probably won't see development of ocean hydro projects. So I think we do need to clarify the respective roles to the agencies.

Senator Murkowski. Mr. Chairman, if there's anything that we on this committee can do to help encourage that along I think it would be very, very beneficial for all. Thank you.

The Chairman. Thank you very much.

Senator Menendez.

Senator Menendez. Thank you, Mr. Chairman. I appreciate your leadership here on the committee on this particular issue. It seems to me that we need action that matches the sense of urgency that we are feeling on climate change.

We've heard testimony here and elsewhere that China is building a coal fired plant every week. It seems to me if we have any hope of lowering greenhouse gas emissions and avoiding catastrophic climate change we need to act quickly and effectively. But it also seems to me if we want technologies like carbon capture and sequestration to flourish in the near future it would appear to me that we need to pass a cap and trade bill as soon as possible because when carbon emissions have a price attached to them well the coal powered industry will act quickly by investing their own money and not just relying on Federal research.

I was listening to Senator Dorgan's numbers as he was putting them out there and other numbers that we've heard projected. We're talking about an enormous amount of money. I think there's a good part of who participates in this process because while we need to act quickly, I think we also need to act wisely.

The issue I'd like to pursue with this panel to some degree is I think maybe one of the most important issues in the question of the regulatory regime and that's the question of who pays to over-see and regulate this new effort. I mean I look at Future Gen which some of us have believed is has always been Never Gen. I look at the statement made by the Deputy Energy Secretary, Clay Sell, who said among the reasons why they were dropping and this was just as in December the Department was listing this as the centerpiece of their strategy for clean coal technologies. One of the major reasons is that the price had risen to 1.8 billion dollars.

That I think goes to the very heart of the question of who participates in this process in paying toward it. You know we have had 150 years of electric power and fossil fuel that has, yes, it has lit up America, but it's also caused some significant environmental damage along the way and a lot of public health concerns including acid rain, mercury poisoning, asthma attacks, ozone depletion, particulate matter pollution, just to name a few. There are no apologies, no apologies for any of this.

Instead now when the industry is under threat they want the American taxpayers to save them. I think it has it backward. I'm wondering whether one of the views that we should have is that in fact an industry that is mature and immensely profitable shouldn't be significantly in the forefront of the concept of polluter pays for example is one.

So my question that I wanted to throw open to the panel is if we're likely to have to ensure CO_2 stays in the ground for 100 years or more who's going to ensure. Who's going to pay for that process of making sure that we have the pipelines, that we have the monitoring, that we in essence are going to ensure that something, that we're going to pursue that course on is going to be one that is not born specifically by the American taxpayer. Do we view this being born by the American taxpayer? You don't all have to jump to answer the question.

Mr. Grumbles. Just from an EPA perspective we understand that is one of the key questions and as we're moving forward on the regulatory piece that relates to safeguards for underground sources of drinking water. As I mentioned earlier the financial responsibility question for not only the closure, but the post closure and monitoring is one that we're going to be getting comments from the public on in the rulemaking process. Outside of the rulemaking process, we have an established workgroup with DOE and with States.

A key question on that is also the important role of the States in this effort. From the Safe Drinking Water Act standpoint, the UIC program, 35 States have been delegated the authority to run the program and many States are very interested in carbon sequestration, the new frontier. We're going to be working with States on that question of liability.

EPA continues to embrace the polluter pays principle. When it comes to government oversight and management that's going to be one of the key issues; post closure monitoring and financial responsibility.

Senator Menendez. Many of us find the EPA weakening the provisions of polluter pays by virtue of the fact that Superfund is a perfect example of how we have not met that standard. But my question is to the rest of you. You know we're talking about an enormous outlay of pipelines, of the sequestration caverns which we will put this in, the monitoring for the leaking and just the technology. Isn't there a participation of some significant degree by the industry or do we see the American taxpayer putting this all out there and ultimately not having a very robust, to say the least, participation by the industry at the end of the day.

Mr. Slutz.

Mr. Slutz. We really how this has developed just like the rest of the infrastructure of this country will be by the private sector and the market and in the case of DOE technologies the idea is to create these technologies so the market can pick them up. One of the issues with reference to the Future Gen project that was 74 percent U.S. Government funded and 26 percent industry. The revised approach that we have come out with while we're still we have a request for information. We're seeking feedback, but would likely the way this comes out would reverse that percentage and have a much larger private sector share in a power plant than it would be much more commercially oriented. So it is built into that.

I would add one more thing as we deal with issues of carbon. One of the challenges which I know all of you are aware of is how do you develop the technology. How do you implement them without dramatically increasing the price of energy because as you say that will then impact our economy and impact citizens.

Senator Menendez. Mr. Chairman I know my time is over, but, you know, I find it interesting just two notes here for future reference. Powers affect the price of energy, of course it's important. But by the same token if the government is funding overwhelmingly all of this effort and then the monitoring, that is the taxpayer too. They may not see it in their energy bill, but they're seeing it out of their money in enormous fashion, No. 1.

No. 2, I find it interesting we had a hearing on a different matter, Mr. Chairman on the Banking committee where the whole issue is what are we doing to try and help people of this country having the American dream not become the American nightmare. A lot of those who say don't get engaged. You know, let the market work its forces.

If we're not going to solve the problem for people who ultimately may be losing their homes was the centerpiece of the American dream, then I'm not so sure the American taxpayer can be called upon to solve the problems for the coal industry, be responsible, overwhelmingly for the problems of the coal industry in trying to meet the challenges of the future. I think there has to be some symmetry at the end of the day as to what's our views in terms of responsibility. But, thank you, Mr. Chairman. I appreciate the opportunity.

The Chairman. Thank you very much. We have another panel right after this. Let me ask if Senator Corker has any final question to put to this panel?

Senator Corker. Mr. Chairman, thank you and Senator Menendez, I agree with you. I think this is what we need to get away from is any subsidizing. We ought to figure out the true cost of energy and I hope it will be industry funded.

We're all going to be looking at bills. Senator Bingaman has a very thoughtful bill. There's other bills regarding cap and trade. Regardless of whether the money is there from the private sector to put carbon in the ground or not, the first thing we all have to be comfortable with is it is safe to do that. I mean can we, are we really solving a problem? Are we not having other unintended problems occur?

I'm just wondering if you all could give us a guess, if you will. We got a lot of departments that work on this. It's kind of a cluster making everything happen sequentially. Do you all have any idea when we're going to know for real, in a way that we can really pump some significant resources toward this from this from the private sector, when we're really going to know when it is safe and we have the regulations in place? It would be helpful to us.

We're going to have to be dealing with credits and allowances and all those kinds of things. Just to have an idea of when you think on a mass basis we will be able to do that? I would just love a number.

Mr. Grumbles. From an EPA standpoint the key part of answering that question is getting in place the safeguards up front. We believe that there's very minimal risk associated with carbon sequestration if you do have the proper geologic siting and well construction and monitoring and post closure monitoring. We're learning a lot by the work that is going on around the world in some of the demonstration sites. So we're optimistic about this promising, but still unproven technology.

IPCC basically has said it's very promising from their perspective. They say that carbon dioxide could be trapped for millions of years in appropriate geologic formations are likely to retain over 99 percent of the injected CO_2 over 1,000 years. So it is a question of when you go from the smaller and experimental demonstration sites to the larger commercial scale sites. Will you have the basic safeguards in place to ensure the proper siting and monitoring and measuring? If so, then we're very optimistic about the safety of this long term storage.

Senator Corker. That seems vague to me.

The Chairman. Does anybody else want to respond?

Mr. Slutz. I will respond. We can get some more information with some of the dates from our—we have a carbon sequestration research road map that lays out certain milestone dates. Rather than throw them out and be incorrect. But we can show some of those things.

One of the challenges is we have a greatly differing and varying geology in this country. That's why we have these seven partnerships to try to get information in different regions, different types of geology. One of the challenges we don't always know, you know, it's research. You don't know exactly what you're going to find. But I think we can put together some information that will show you some of the key milestones on where we're trying to get with our research program and our portfolio to help understandthat.

I don't know if it will, it won't give you that specific date, a year or something. But it'll show you the various milestones that we're getting over the next 5 to 8 years.

[The information referred to follows:]

Following is the response to information (milestones/dates) as to when a system will be in place that assures the safety of carbon sequestration and a regulatory framework to allow ongoing mass storage of CO_2.

DOE agrees with the testimony presented by the Environmental Protection Agency (EPA) on January 31, 2008, which stated as follows:

In March 2007, EPA issued technical guidance to help State and EPA Regionalmanagers in processing permit applications for geological sequestration (GS) demonstration projects under the general UIC regulations.

Under the SDWA [Safe Drinking Water Act], EPA develops minimum requirements for state UIC [Underground Injection Control] programs. States may develop their own regulations for injection wells in their State. These requirements must be at least as stringent as the federal requirements (and may be more stringent). Annually, billions of gallons of fluids are injected underground through wells authorized under State and Federal UIC Programs. This includes approximately 35 million tons of carbon dioxide that are injected for the purposes of enhancing oil and gas recovery. EPA's proposed regulations will build on the UIC Program's many years of experience in safely injecting fluids, including carbon dioxide, into the subsurface.

The proposed regulation, currently in development under an accelerated schedule, will take into account the EPA's existing UIC program requirements. Key components of the proposed regulation will include requirements related to: (1) geologic site characterization to ensure that wells are sited in suitable areas to limit the potential for migration of injected and formation fluids into an underground source of drinking water; (2) well integrity testing and monitoring to ensure that the wells perform as designed; and, (4) well closure, post-closure care, and financial responsibility to ensure proper plugging and abandonment of the injection well. We will also discuss long-term liability and seek further comment on this issue as part of the proposed rulemaking.

The Chairman. Anyone else?

Mr. Allred.

Mr. Allred. Mr. Chairman, Senator Corker. Just a point that I think that there is already a lot of information that is available that would show that it can be done. I think we have to be more deliberate about how that information is collected, but as Mr. Slutz indicated, they have projects where in excess of a million tons of CO_2 have been injected. We have a number of places where over a million tons a year has been injected for enhanced oil recovery.

So I think there are a lot of those things which either are now being done or could be done. The purpose of those, at least the oil recovery projects, has not been sequestration, although there may have been a significant amount of sequestration occur. One of the things I think that we have the potential to do is to add a purpose to those, not to eliminate the other purpose, but add a purpose to those. Answer a lot of questions that you were just asking. I suspect that those answers will be with proper knowledge and proper consideration that we will be safe in can be done fairly quickly.

The Chairman. Thank you. Senator Lincoln, did you have questions of this panel?

Senator Lincoln. No.

The Chairman. Let me thank all the witnesses of this panel and we will stay in touch with you and try to continue moving ahead on this set of issues.

Let me call the final panel forward. Lawrence Bengal who is with the Arkansas Oil and Gas Commission. Tracy Evans who is with Denbury Resources in Plano, Texas and Scott Anderson with Environmental Defense from Austin, Texas. Thank you all for coming.

Yes. Let me call on Senator Lincoln to make whatever introductions she would like of our witnesses here. I know one of these witnesses is from her home State. Blanche, go right ahead with whatever you would like to say.

Senator Lincoln. Mr. Chairman, I apologize that I've been absent for the earlier part of the hearing. As most of us know it gets over scheduled way too much, but I certainly appreciate your leadership in this area. There's so much to be done and so much for us to learn without a doubt.

But it is my pleasure to introduce the Chairman of the Arkansas Oil and Gas Commission, Larry Bengal. Mr. Bengal has held positions as researcher with the Illinois State Geological Survey as a project manager for Geologic and Mining Engineering Consulting firm, engaged in projects throughout the United States and as an independent counseling or consulting petroleum geologists as manager of the Illinois Class II underground injection control program and as a supervisor of the Illinois Oil and Gas division. He is here today in his capacity as chairman of the IOGC carbon capture and geologic storage task force. We're certainly appreciative of all of what he has done.

I feel, like you, Larry, we appreciate you being here and appreciate all the both evidence and intelligence that you bring to the issue that we're dealing with here and grateful that you've joined us. We look forward to continuing to work with you as well. But we're very proud of him in Arkansas and his work in the oil and gas issues there and equally proud of his fine work with the CCGS task force. So thank you, Mr. Chairman and thank you, Larry for being here and to all the panelists.

The Chairman. Alright. Thank you very much. Mr. Bengal, why don't you start and then Mr. Anderson and then Mr. Evans.

Statement of Lawrence E. Bengal, Director, Arkansas Oil and Gas Commission, Little Rock, AR

Mr. Bengal. Good afternoon and thank you, Senator Lincoln for that gracious introduction.

Mr. Chairman, committee members, my name is Lawrence Bengal. I am the director of the Arkansas Oil and Gas Commission and I'm appearing here today in my capacity as chairman of the Interstate Oil and Gas Compact Commissions Task Force on Carbon Capture and Geologic Storage. I would like to share with the committee the experience and conclusions of the task force and offer brief comments on S. 2144 and S. 2323.

Funded by the U.S. Department of Energy through the National Energy Technology Laboratory, the task force has been engaged since 2003 in a two phase effort relating to the regulation of geologic storage of carbon. In phase one the task force undertook a thorough review of the technology of geologic storage. In phase two developed model regulations. A major conclusion of the task force in phase one was that the geologic storage of carbon dioxide or CO_2 was not something entirely new or mysterious, but the technological outgrowth of analogs with which the States already have regulatory experiences.

In phase two the task force has produced a clear and comprehensive model regulatory regime for the geologic storage of CO_2. Utilizing these model frameworks, States and provinces and indeed other nations can begin immediately the process of enacting legislation and promulgating rules and regulations enabling CO_2 geologic storage projects. In fact a number of States have already begun this process.

By 2010 I fully expect that at least 5, 10 or more States will have legal and regulatory systems in place. The EPA carbon storage regulations under the Safe Drinking Water Act and it's implementing underground injection control or UIC programs should also be in place within this timeframe. It is my expectation the State regulatory systems will work seamlessly like hand in glove with the regulations likely to emerge from the EPA regulatory development process. This is largely because of the role States play in the administration of UIC programs under EPA privacy authority.

As concerns SB, Senate bill 2323 the legal and regulatory framework proposed by the task force does not require in order for our program to work effectively any broader, overarching Federal regulation. Framework proposed by the IOGCC task force is comprehensive and contains many aspects that are solely a function of State law. Our expectation is that the combination of State and EPA UIC regulatory systems will produce a flexible, responsive, safe and environmentally sound regulatory framework for CO_2 geologic storage that will be more than adequate to get the first project planned and safely off the ground.

We would suggest that before we rush to create a potentially unnecessary Federal and regulatory framework that we observe how this combined State and EPA administered storage framework functions. If a need for additional Federal regulatory authority manifests, it can be addressed at that time. As concerns SB 2144, I would only suggest that the departments and agencies designated in the bill to conduct the study be required to conduct it in close cooperation and consultation with the States which have much experience in this area.

Let me now turn to the diagram which you see before you which illustrates the cradle to grave regulatory model that was developed by the task force and this is what we recommended to States. There are three phases you can see. Although I do not have enough time at this stage to go over each of the phases, this will give you some idea of the breadth of the regulatory structure proposed by the IOGCC task force.

I would note however that only within the project areas indicated by the green box does it appear to the EPA has regulatory authority over the Safe Drinking Water Act. The other areas would be covered under State law.

Let me close by noting the obvious that public support for carbon storage as a strategy for mitigating the impact of global climate change will be crucial. It is important to educate the public about this technology including CO_2 long history of being transported, handled and used in a variety of applications. CO_2 is certainly no more, if not less, than the hazardous waste of natural gas and calling it such makes it very difficult for public acceptance of CO_2 storage.

It will also be vitally important to include the public in every step of the regulatory development process at the State and Federal levels. State laws will ensure public notice and participation and the State processes of both legislation and regulation development stages. Thank you for the opportunity to appear here today. If I can provide any information, please do not hesitate to ask. I would ask though that a copy of the full IOGCC task force be included in the record today.***

[The prepared statement of Mr. Bengal follows:]

Prepared Statement of Lawrence E. Bengal, Director, Arkansas Oil and Gas Commission, Little Rock, AR

Good afternoon. My name is Lawrence Bengal. I am the Director of the Arkansas Oil and Gas Commission and I'm appearing here today in my capacity as Chairman of the Interstate Oil and Gas Compact Commission's Task Force on Carbon Capture and Geologic Storage (CCGS). The Task Force was comprised of representatives from IOGCC member state and provincial oil and gas agencies, U.S. Department of Energy sponsored Regional Carbon Sequestration Partnerships, the Association of American State Geologists, industry experts, as well as representatives from the U.S. Environmental Protection Agency (EPA), the U.S. Bureau of Land Management (BLM) and the environmental group, Environmental Defense, who attended as observers.

The member states of the Interstate Oil and Gas Compact Commission (IOGCC) produce more than 99% of the oil and natural gas produced onshore in the United States. Formed by Governors in 1935, the IOGCC is a congressionally ratified inter-state compact. The organization, the nation's leading advocate for conservation and wise development of domestic petroleum resources, includes 30 member states, associate states, and 4 international affiliate provinces. The mission of the IOGCC is two-fold: to conserve our nation's oil and gas resources and to protect human health and the environment during the production process. Our current chairman is Governor Sarah Palin of Alaska.

I am here today to share with the committee the experience and conclusions of IOGCC's CCGS Task Force and to offer our comments on S. 2144, the ''Carbon Dioxide Pipeline Study Act of 2007'', and S. 2323, the ''Carbon Capture and Storage Technology Act of 2007''.

Funded by the U.S. Department of Energy (DOE) and its National Energy Technology Laboratory (NETL), the Task Force has been engaged since 2003 in a twophase effort relating to the regulation of the geologic storage of carbon. In Phase I, the Task Force undertook a thorough review of the technology of geologic storage and in Phase II developed a model statute and model rules and regulations for the states and provinces to administer regulatory oversight of geologic storage of carbon dioxide (CO_2).

A major conclusion of the Task Force in Phase I was that the geologic storage of CO_2, in addition to conservation, is among the most immediate and viable strategies available for mitigating the release of CO_2 into the atmosphere. It was readily apparent to the Task Force that carbon storage was also not something entirely new and mysterious—but the technological outgrowth of four analogues. These four analogues, in the opinion of the Task Force, provide the technological and regulatory basis for storage of CO_2 in geologic media: 1) naturally occurring CO_2 contained in geologic reservoirs, including natural gas reservoirs; 2) the large number of projects where CO_2 has been injected into underground formations for Enhanced Oil Recovery (EOR) operations; 3) storage of natural gas in geologic reservoirs; and 4) injection of acid gas (a combination of H2S and CO_2), into underground formations, with its long history of safe operations.

It was the opinion of the Task Force that given the jurisdiction, experience, and expertise of the states and provinces in the regulation of oil and natural gas production as well in regulating the analogues identified above, the states and provinces would not only be well able to regulate, but would be the most logical and experienced regulators of CO_2 geologic storage. Additionally and importantly, the oil and natural gas producing states and provinces are strategically and geologically wellsituated for the geologic storage of CO_2. Regulations already exist in most oil and natural gas producing states and provinces covering many of the same issues that need to be addressed in the regulation of CO_2 geologic storage, and consequently serve as adaptable frameworks.

Given these Phase I conclusions, the Task Force, in Phase II, began work and in September of 2007 produced, for the first time, a clear and comprehensive model legal and regulatory regime for the geologic storage of CO_2. Utilizing these model regulatory frameworks, states and provinces, and indeed other nations, can begin immediately the process of developing and enacting legislation and promulgating rules and regulations enabling CO_2 geologic storage projects. California, New Mexico, North Dakota, Texas, and Wyoming are, among other states, in various stages of developing such a legal and regulatory framework.

I anticipate that by 2010 there will be at least 5–15 states, encompassing much of the country best suited for carbon geologic storage, with legal and regulatory systems in place for the regulation of geologic storage of CO_2. I would also anticipate that in this same general timeframe that EPA will likewise have in place regulations governing geologic storage of CO_2 under the Safe Drinking Water Act and the implementing Underground Injection Control (UIC) Program.

It is appropriate that I now briefly address how the IOGCC anticipates the EPA's CO_2 geologic storage regulations will interface with the regulatory systems being developed by the states. Given the incorporation of UIC-like regulatory requirements into the proposed IOGCC model regulatory frameworks, there is every reason to anticipate that the IOGCC and EPA frameworks will fit like hand in glove. This is largely because of the role that states play in the administration of UIC programs under EPA primacy authority.

In this regard, as you are no doubt aware, the EPA is in the process of developing regulations for geologic sequestration under the Safe Drinking Water Act with the goal of having draft regulations for public comment by the summer of 2008. The IOGCC at the invitation of EPA has two representatives, Berry ''Nick'' Tew of Alabama and myself, actively participating in the process as state co-regulators. States with primacy already play an integral role in administering the UIC program and under future rules governing geologic storage, are likely to do so again. Having representatives from states involved in the process helps insure compatibility between the state and federal components of geologic storage regulatory oversight.

What is clear to me, especially given my involvement with the current EPA workgroup, is that the state regulatory system for carbon storage proposed by the IOGCC Task Force will in all likelihood work seamlessly with the regulations likelyto emerge out of the EPA regulatory development process.

Having made this observation let me now offer brief comment on the two bills which are the subject of this hearing today.

The legal and regulatory framework proposed by the Task Force, which I will discuss in more detail subsequently, does not require, in order to work effectively, broader over-arching federal regulation such as that apparently contemplated by S.2323. The Task Force strongly believed that what it was proposing was comprehensive and given the number of aspects that are solely functions of state and not federal law (for example ownership of storage rights and means for amalgamating those rights through some type of condemnation proceeding) that there was no need or in some respects even the possibility of a broader federal role. It is suggested that there will be ample time over the coming years to see how the state-administered CO_2 storage frameworks function in tandem with the EPA UIC storage regulations thereby alleviating the need to rush to create a potentially unnecessary federal regulatory framework at this time. If there is need for additional federal regulatory authority, it can be addressed legislatively then. I fully anticipate that what will exist during this interim period will be a flexible, responsive, and environmentally sound combination of state and EPA UIC regulatory systems, which will be more than adequate to get the first projects planned and off of the ground. Experience with these projects will show us rather quickly if weaknesses or problems with the existing frameworks manifest. We have absolutely no expectations at this time that they will.

As concerns S. 2144 and the requirement of a study of feasibility relating to construction and operation of pipelines and CO_2 sequestration facilities, I would notefirst that the Task Force's proposed legal and regulatory infrastructure encompasses construction and operation of CO_2 sequestration facilities. Second, I would suggest that given this, that the federal government departments and agencies designated in the bill to conduct the study at a minimum be required to conduct the study in close cooperation and consultation with states, including representatives of the IOGCC Task Force.

It is now appropriate to supply a little more detail about the legal and regulatory system which the IOGCC Task Force has proposed for the geologic storage of CO_2 and how, precisely, the proposed EPA regulatory system for CO_2 storage would likely fit into this system. This ****diagram will be helpful:

The diagram represents the ''cradle to grave'' regulatory model which the Task Force has recommended to states. There are three phases.

1. Licensing Including Amalgamation of Storage Rights

The first phase is the licensing phase which includes the critical requirement that the project operator control the storage rights.

The Task Force concluded that as a part of the initial licensing of a storage project that the operator of the project must control the reservoir and associated pore space to be used for CO_2 storage. The operator would need to acquire theserights from the owners or assume those rights by means of eminent domain, unitization or some other vehicle that either exists in a state or would be created by the state uniquely for this purpose. This step is necessary because in the U.S., the right to use reservoirs and associated pore space is considered a private property right and must be acquired from the owner. It was the conclusion of a Task Force legal subgroup that in most U.S. states, for non EOR-related storage, the owner of these rights would likely be the owner of the surface estate. It may be prudent, however, depending upon the specific property right ownership framework in a given state, for an operator to also control the relevant subsurface mineral rights.

Additionally, as part of the initial licensing of a project the operator would be required to submit for State Regulatory Authority (SRA) approval, detailed engineering and geological data along with a CO_2 injection plan that includes a descriptionof mechanisms of geologic confinement that would prevent horizontal or vertical migration of CO_2 beyond the proposed storage reservoir. The operator would also be required to submit for approval by the SRA a public health and safety and emergency response plan, worker safety plan, corrosion monitoring and prevention plan and a facility and storage reservoir leak detection and monitoring plan.

The rules also include requirements for an operational bond that would be sufficient to cover all operational aspects of the storage facility excluding wells which would be separately bonded.

Site licensing and amalgamation of storage rights is generally believed to be outside the scope of the current UIC Program, and given that regulatory involvement with property rights is a state issue, this phase is best addressed at the state level. In addition, given the likely competition for acceptable storage sites, it is in a state's interest to manage these sites to maximize storage capacity and resolve any operator conflicts over the right to use storage resources, thereby maximizing the state's best economic interest in providing storage sites for that state's generators.

2. The Storage and Closure Phase

In this second phase we are talking about the phase, following initial licensing, when the storage project is developed, operated, and closed. This includes a short time period following plugging of the wells during which time the project is monitored to ensure stability of the injected CO_2.

During the storage component of this phase the model rules specify the procedures for permitting and operating the project injection wells to safeguard life, health, property and the environment. The operator would be required to post individual well bonds sufficient to cover well plugging and abandonment, CO_2 injection and/or subsurface observation well remediation. The rules also specify design standards to ensure that injection wells are constructed to prevent the migration of CO_2 into other than the intended injection zone. Provisions in the rules also ensure that all project operational standards and plans submitted during the licensing phase would be adhered to and that the project and wells are operated in accordance with all required operating parameters and procedures. Quarterly and annual reports would be required throughout the operational life of the project. The rules also ensure that the wells are properly plugged and the site restored. The individual well bonds, maintained during the operational phase of the project would be released as the wells are plugged.

The closure component of this phase is defined as that period of time when the plugging of the injection wells has been completed and continuing for a defined period of time (10 years unless otherwise designated by the State Regulatory Authority) after injection activities cease and the injections wells are plugged. During this closure period, the operator of the storage site would be responsible for providing the required data to ensure the injected CO_2 has not migrated beyond the project boundaries and the injected CO_2 plume has been stabilized. During this time the operator is required to maintain an overall project operational bond.

This phase is primarily where EPA is developing proposed rules to ensure the operation and plugging of the wells are protective of the groundwater resources under the UIC Program.

3. Long-Term ''Care Taker'' Phase (long-term monitoring and liability)

The last phase is the Long Term or Post-Closure Period and is referred to as that period of time when the operator of the project is no longer the responsible party and the long-term ''care taker'' role is assumed by a government entity or government-administered entity. The major issue faced by the Task Force was how to deal with long-term monitoring and liability issues. The formula settled upon by the Task Force is the following:

At the conclusion of the Closure Period, the operational bond would be released and the regulatory liability for ensuring that the site remains a secure storage site would transfer to a trust fund administered by the state. During the Post-Closure Period, the financial resources necessary for the state or a state-contracted entity to engage in future monitoring, verification, and remediation activities would be provided by this state-administered trust fund.

The Task Force concluded that such a state-administered trust fund would be the most effective and responsive ''care-taker'' to provide the necessary oversight during the Post-Closure Period. The trust fund would be funded by an injection fee assessed to the site operator and calculated on a per-ton basis.

In summary, the EPA Regulations under the SDWA and the UIC Program primarily deal with the Storage and Closure Phase as illustrated by the green box in the diagram, for it is only in the project areas within that box that EPA has authority under the SDWA. In addition to EPA's mandate to protect drinking water under the SDWA, the IOGCC regulations cover other public health and safety issues that need to be a part of a comprehensive regulatory framework. As previously stated, almost all of the well operational standards proposed in the IOGCC model regulations are already UIC requirements of one form or another.

What I anticipate is that the proposed EPA regulations, whatever they end up being, will yield a set of uniform national standards, which superimposed on whatever state regulations may be in place will result in national consistency of application so as to ensure that drinking water resources are protected. Again as previously stated, given most states (those with primacy) already administer the existing UIC program, they will continue to do so, conforming their state regulations as they per-tain to the geologic storage of carbon to the minimum standard set by the new EPA regulations.

Unless the EPA regulations end up being unnecessarily proscriptive and onerous, the systems should work together perfectly and as I've already stated, ''seamlessly''. Certainly this is the hope and current full expectation of the IOGCC.

I will note that with regard to federal lands (surface and/or mineral interests), that federal regulations emanating out of the BLM will undoubtedly be necessary. However, what emanates out of BLM would in all likelihood be more akin to what the states have done with regard to state and private lands rather than an over-arching and broader national regulatory scheme.

Additionally, our model regulatory system does not address the regulatory issues involving CO_2 emissions trading and accreditation for the purpose of securing carbon credits. The Task Force concluded that the issue of CO_2 emissions trading and accreditation would likely best be addressed in the marketplace and/or at the federal government level and was beyond the scope of the Task Force's mandate. In any event, the Task Force strongly believes that development of any future CO_2 emissions trading and accreditation regulatory frameworks should utilize the experiences of the states.

As concerns long term ''care taker'' liability, what the Task Force has proposed will have to be addressed by each state and province as they develop their own framework. It remains to be seen if states will agree with the Task Force or propose something new. There may indeed be a need for a federal role here at some point in the future but it is suggested that federal action in this area await a clear need manifesting itself in the years ahead.

Additionally and very importantly, states and provinces are likely to continue to regard CO_2 geologic storage reservoirs as a valuable resource that should be managed using resource management frameworks, therefore avoiding the treatment of CO_2 storage as waste disposal. The Task Force strongly believes that treatment of CO_2 as a waste under waste management regulatory frameworks will diminish significantly the potential of carbon storage technology to meaningfully mitigate the impact of CO_2 emissions on the global climate. The energy consuming public and the industry which produces that energy share a common goal in coming up with a workable solution.

Let me close by noting the obvious—that public support for carbon storage as a strategy for mitigating the impact of global climate change will be crucial. It will be important to educate the public about this technology including CO_2's long history of being transported, handled, and used in a variety of applications. It will also be vitally important to include the public in every step of the regulatory development process, state and federal. State open meeting laws will ensure public notice and participation in the state process at both legislation and regulation development stages.

Thank you for the opportunity to appear here today. If I can provide any additional information, please do not hesitate to ask.

The Chairman. We'll be glad to include that in the record. Thank you very much for your testimony.

Mr. Anderson.

STATEMENT OF SCOTT ANDERSON, ENVIRONMENTAL DEFENSE, AUSTIN, TX

Mr. Anderson. Thank you, Mr. Chairman. We're pleased to be here today as the committee considers how to create a regulatory framework regarding carbon capture and storage so that CCS can play a role in the fight against climate change. Climate change is the most significant environmental issue of our generation. The Senate is doing important work in this area. Cap and trade legislation if adopted would do a lot to commercialize CCS by creating a market value for voiding CO_2 emissions and considering the measures you have before you today it is vital work as well.

Without a sound regulatory framework uncertainty will prevail and the marketplace will not be able to move CCS forward in a significant way. Public acceptance will happen only if the public is confident that rigorous and credible regulatory oversight is in place. The fact that Environmental Defense supports deployment of CCS does not mean that we are champions of coal.

We believe that business as usual for coal is over. Public opinion is shifting and conventional coal plants are being delayed or canceled at a rate unimaginable a year ago. People are increasingly recognizing that energy efficiency and renewables should play a leading role in energy and climate policy.

We're not champions of coal at Environmental Defense but we are realists. Coal will continue to be used for the foreseeable future and we believe that CCS can play a significant role in helping coal to reduce its greenhouse emissions. Even today in the absence of a full fledged private market it's possible where the economics arrive to begin deployment.

The Texas legislature passed a bill in 2007 that provided a severance tax incentive for oil producers who use CO_2 to produce oil and then sequester the carbon afterwards, defining permits as meaning 99 percent retention for 1,000 years or more. So at least in Texas the legislature has made a determination that CCS is ready for deployment now.

I'll turn now to Senate bill 2144 and section 5 of Senate bill 2323. Senator Coleman's Senate bill 2144 would require a feasibility study that we believe is sound. We endorse this measure.

Section 5 of Senator Kerry's bill would establish an interagency task force to develop regulations and we believe that with some modification this is worthy of passage as well. Section 5 has several notable strengths. It assures that the development of a regulatory framework will move forward expeditiously that includes the Departments of Energy and Interior in the process. It appropriately names the administrator of the EPA who has key responsibilities of the Safe Drinking Water Act to be chairman or chairperson of the task force.

Finally the legislation builds on existing regulatory authority on an incremental as needed bases. Subsection (a)(5) requires regulations to take into account the existing UIC program and then continues to provide additional requirements that regulations must satisfy. We believe such a step by step approach is prudent for first generation CCS rules. As the need for additional grants of jurisdiction or congressional guidance become apparent additional provisions can be enacted through supplemental legislation.

There are also several areas of Section 5 that we feel could be improved. I'll only touch verbally on one of those. We are confident that the bill is intended to accelerate adoption of carbon sequestration regulations, but EPA is already engaged in rulemaking and there's a risk that the bill can actually slow down adoption of the first set of regulations. We recommend adding a provision indicating that Congress does not intend to discourage rulemaking in the near term, but rather intends that regulations should reflect the interagency process spelled out in the bill. If EPA adopts rules based on existing procedures in the meantime in a rules developed pursuant to this bill would become the second generation of rules.

The final portion of my prepared testimony discusses the appropriate design of geologic sequestration regulations. We suggest that rules generally should be flexible and performance based and that they should adapt to evolving knowledge and best practices. At the same time we say that it's not enough to be flexible, adaptive and performance based. It's essential that rules be grounded in a thorough scientific understanding of the risks involved and the rules assure that the risk will be managed properly. Some aspects of the rules such as site characterization and selection requirements will need to be relatively more prescriptive than others. With that I'll close. Thank you.

[The prepared statement of Mr. Anderson follows:]

Prepared Statement of Scott Anderson, Environmental Defense, Austin, TX

We appreciate the opportunity to speak to you today as the committee considers how to create a regulatory framework that will enable carbon capture and storage (CCS) to play a role in the fight against climate change. Climate change is the most important environmental issue of our generation and successful development anddeployment of CCS is a critical path for taking coal, the world's most abundant but carbon-intensive fossil fuel, and accommodating it to a carbon-constrained future.

Environmental Defense is a national non-profit organization representing more than 500,000 members. Since 1967, we have linked science, economics and law to create innovative, equitable and cost-effective solutions to urgent environmental problems. My personal background includes more than 20 years representing inde-pendent oil and gas producers in Texas, and so I have some appreciation for many of the issues and concerns related to the underground storage of carbon dioxide.

The Senate is doing important work to address the threat of climate change. The single most important thing the Senate can do to commercialize CCS is to take quick action on cap and trade legislation, since such legislation would create a mar-ket value for avoiding carbon dioxide emissions. Given the right incentives, we believe that the market will be far more effective and efficient in discovering necessary technologies of all types, including CCS, than any suite of government mandates or subsidies, however well intentioned.

Consideration of regulatory measures such as those before you today is vital work as well. Without a sound regulatory framework to govern carbon capture, transportation and storage, uncertainty will prevail and the marketplace will not be able toachieve the kind of deep and sustained reductions necessary to avoid the worst consequences of greenhouse gas buildup. Similarly, public acceptance of CCS will happen only if the public is confident that rigorous and credible regulatory oversight is in place.

The fact that Environmental Defense supports the deployment of CCS does not mean that we are champions of coal. We believe that business as usual for coal isover. Public opinion is shifting and conventional coal plants are being delayed or canceled at a rate unimaginable even a year ago. In states like Texas, Florida, Oklahoma and Kansas, people are beginning to realize that it is environmentally irresponsible and fiscally imprudent to proceed with building new coal plants, absent a concrete plan to reduce and avoid CO_2 emissions. We are also pleased that people are increasingly recognizing that energy efficiency and renewables should play aleading role in energy and climate policy.

Although we are not champions of coal at Environmental Defense, we are realists. Coal will continue to be used for electricity production for the foreseeable future. Therefore the nation and the world need technologies that enable coal to be used in a manner that avoids significant greenhouse gas emissions. According to an IEA study released in 2006, CCS could rank, by 2050, second only to energy efficiencyas a greenhouse gas control measure. The Intergovernmental Panel on Climate Change projects that CCS could, by 2100, contribute 15 to 55% of the greenhouse gas reductions needed to avert catastrophic climate change. Just last week in a proposed directive on CCS, the Commission of the European Communities noted that efficiency and renewables are the most sustainable supply options in the long run but that ‘‘we cannot reduce EU or world CO_2 emissions by 50% in 2050 if we do not also capture CO_2 from industrial installations and store it in geological formations.’’

While different analysts come up with somewhat different scenarios, it is clear that coal is not going to disappear anytime soon and therefore effectively capturing and sequestering carbon dioxide emissions from coal can make a real difference in whether mankind will be able to solve climate change problems. We are fortunate that early sequestration projects, together with over 30 years of experience with injecting CO_2 into oilfields, have provided confidence that long-term sequestration in properly selected geologic formations is feasible.

In fact, even today, when large-scale commercialization of CCS is hampered by the absence of price signals that could be provided by a market in trading allowances, it is possible to begin deployment and start making real reductions in CO_2 emissions. McKinsey & Company's recent study, ''Reducing U.S. Greenhouse Gas Emissions: How Much at What Cost?,'' provides a sense of the costs involved. My fellow panelist Tracy Evans of Denbury Resources can speak from direct experience about the feasibility of deploying CCS in the oilfield context.

Summary of Comments on S. 2144 and S. 2323

I would like to cover several things this morning. I will touch briefly on S. 2144, which would require a study of the feasibility of constructing and operating carbon dioxide pipeline and sequestration facilities. I want to focus most of our remarks, however, on Section 5 of S. 2323, which would establish an interagency task force to develop regulations for CO_2 capture and storage. Our remarks on Section 5 will focus on regulations for geologic sequestration, rather than capture. Finally, we will offer comments on the appropriate design of sequestration regulations. We will mention why it is important for CO_2 storage regulations, especially in the early years, to be relatively performance-based rather than prescriptive and why it is important for the regulatory framework to adapt as knowledge improves.

We believe that it would be useful to adopt S. 2144, and Section 5 of S. 2323 if modified in several respects, as stand-alone measures. These measures would be most useful, however, if enacted as part of or in concert with comprehensive cap and trade legislation that would create a market value for avoiding CO 2 emissions and thereby encourage market participants to engage in the activities that these measures are intended to address.

S. 2144

Senator Coleman's S. 2144 would require the Secretary of Energy, in coordination with certain other agencies, to study the feasibility of constructing and operating carbon dioxide pipelines and sequestration facilities. We believe that the scope of the contemplated study is sound and that the study is likely to yield important information. Without prejudice to the possibility that others may have valuable suggestions on improving the scope of the study, we generally endorse this bill as proposed.

Section 5 of s. 2323

Section 5 of Senator Kerry's S. 2323 would establish an Interagency Task Force ''to develop regulations providing guidelines and practices for the capture and storage of carbon dioxide.''

Section 5 has several notable strengths:

1. The most fundamental benefit of Section 5 lies in assuring that the development of a regulatory framework for CCS will move forward expeditiously. The intent is clearly that issuance of regulations should be accelerated, not delayed.
2. Including the Departments of Energy and Interior in the regulatory development process is worthwhile. DOE has significant expertise in carbon capture and sequestration that can benefit the rulemaking process. The Department of Interior's Geologic Survey also has significant expertise and is in a position to offer useful input.
3. The bill appropriately names the Administrator of the Environmental Protection Agency as the chairperson of the task force. This is appropriate given that EPA, in addition to having its own significant expertise in CCS, has responsibility under the Safe Drinking Water Act's Underground Injection Control Program to protect underground sources of drinking water from contaminants that might cause a violation of a national primary drinking water regulation or otherwise adversely affect the health of persons.
4. The legislation builds on existing regulatory authority on an incremental, asneeded basis, i.e. subsection (a)(5)(A) requires that the regulations ''take into account existing underground injection control program

requirements'' and then provides additional requirements that regulations must satisfy in subsections (a)(5)(B)-(F). We believe it is prudent to take such a step-by-step approach to authorizing and overseeing the development of ''first generation'' rules for CCS. Both industry and regulators will ''learn while doing'' in the early years of this technology. For now, most observers (ourselves included) appear to find the Safe Drinking Water Act's Underground Injection Control Program to be generally adequate as a basis for initial federal regulations. As the need for additional grants of jurisdiction and/or Congressional guidance becomes apparent, additional provisions can be enacted through supplemental legislation.

There are also several aspects of Section 5 where the committee has an opportunity to make improvements:

1. As noted above, subsection (a)(5)(B)-(F) builds on the Safe Drinking Water Act by requiring that carbon dioxide capture and storage regulations satisfy several objectives that are not part of the existing underground injection control program. However, in our judgment, two more requirements ought to be added. These are (to borrow language from the proposed Lieberman-Warner Climate Security Act): (a) a requirement to regulate the ''long-term storage of carbon dioxide and avoiding, to the maximum extent practicable, any release of carbon dioxide into the atmosphere;'' and (b) a requirement that the carbon dioxide storage regulations protect not just underground sources of drinking water and human health, but ''the environment'' as well. In order to fill these two gaps, we recommend borrowing the language just quoted from section 8001 of S. 2191.
2. We are confident that S. 2323 is intended to accelerate the adoption of carbon sequestration regulations (while at the same time broadening the regulatory development process beyond EPA). There is a risk, however, that the bill could actually slow down adoption of EPA's first set of regulations, which the agency currently plans to propose in the Federal Register by this fall. Publication and adoption of rules in the near term would be likely to have a positive effect on the development of early CCS projects. It would be extremely unfortunate if passage of S. 2323 served to convince EPA to wait for the conclusion of the S. 2323 process before adopting the first set of regulations. Accordingly, we recommend that a provision be added to the bill indicating that Congress does not

intend to discourage earlier CCS rulemaking but rather desires to make sure that regulations growing out of an interagency process are adopted in the near-term. If EPA adopts rules based on existing procedures in the meantime, the regulations developed pursuant to S. 2323 would become the second generation rules.

3. Subsection (a)(5)(C) requires carbon dioxide storage regulations to "address the potential appropriate transfer of liability to governmental entities." We would prefer that any regulations transferring liability to governmental entities be postponed until after the task force report called for in Section 8004 of S. 2191. If such regulations are authorized sooner, however, we think additional guidance is desirable in order to assure that those who develop the regulations recognize that shifting liability to the taxpayers affects the taxpayers differently depending on whether or not monitoring has demonstrated that the storage project in question is performing as expected. The current proposal in Europe regarding the transfer of liability, released January 23 by the Commission of the European Communities, would transfer liability to the government only "if and when all available evidence indicates that the stored CO_2 will be completely contained for the indefinite future." (Proposed Article 18, Proposal for a Directive of the European Parliament and of the Council on the Geological Storage of Carbon Dioxide). Perhaps that would be a good policy for the United States as well. It would protect the taxpayer and assure that project developers maintain an incentive to operate projects safely and effectively. At a minimum, however, we recommend that subsection (a)(5)(C) of Section 5 be amended so that those who draft regulations addressing liability will do so "taking into account whether or not particular projects have demonstrated a reasonable likelihood that virtually all the CO_2 stored will remain sequestered permanently."
4. Subsection (a)(4) of Section 5 calls on the Interagency Task Force to consult with industry, legal and technical experts. We suggest that consultation be expanded to include experts from non-governmental public interest organizations.

Appropriate Design of Geologic Sequestration Regulations

Geologic sequestration of carbon dioxide is feasible under the right conditions. It has been successfully demonstrated in a number of field projects, including several large projects. The Intergovernmental Panel on Climate Change (IPCC) Special Report on Carbon Capture and Storage concluded in 2005 that the fraction of CO_2 retained in ‘‘appropriately selected and managed geological reservoirs’’ is likely to exceed 99% over 1000 years. The IPCC also concluded that the local health, safety and environmental risks of CCS are comparable to the risk of current activities such as natural gas storage, enhanced oil recovery and deep underground storage of acid gas if there is ‘‘appropriate site selection based on available subsurface information, a monitoring programme to detect problems, a regulatory system and the appropriate use of remediation methods to stop or control CO_2 releases if they arise.’’

While there is little doubt that geologic sequestration is feasible, and little doubt that successful projects are technically achievable today, knowledge and understanding are expected to increase dramatically as the technology begins to be deployed on a large scale. Current projects are highly customized. There are gaps in our knowledge and neither government nor industry have yet developed standard protocols for fundamental aspects of the process such as site characterization and monitoring. The IPCC Special Report projects that increasing knowledge and experience will ‘‘reduce uncertainties’’ and ‘‘facilitate decision-making.’’

In other words, we know enough to get started but we can expect to experience a lot of ‘‘learning by doing.’’

What are the implications of this for the regulatory system? We believe at least four recommendations are in order to account for the fact that increasing knowledge and experience will facilitate rational decision-making in different ways over time:

1. Lean toward a performance-based system. ‘‘Performance-based’’ regulations and ‘‘command-and-control’’ regulations do co-exist—they are two poles on a continuum;
2. Be reasonably flexible. Different projects will present different risks and uncertainties, and the uncertainty presented by a single project will tend to decline over time;

3. Require projects to employ an iterative process, informed by monitoring results and perhaps even by experience gained from other projects, in order to reduce uncertainty and drive improvements in site characterization, site suitability assessment, models, model inputs, field operations, the monitoring plan itself, and the remediation plan;
4. Write ‘‘adaptive’’ rules. Look for language that automatically accommodates evolving best practices. Also structure rules to make use of evolving knowledge at each particular site. Be willing to amend rules when needed to protect the environment, giving due regard to the fact that it generally is in the public interest for the regulatory framework to give the regulated community the certainty needed to make investment decisions.

At the same time, it is not enough for rules to be flexible, adaptive and performance-based. It is essential that rules be grounded in a thorough, scientific understanding of the risks involved and that rules assure that the risks will be managed properly. In order to accomplish this, some aspects of the rules (e.g. site characterization and site selection requirements) will need to be more prescriptive than others.

Conclusion

In a carbon-constrained world where market forces are harnessed to make sure that society’s carbon footprint is reduced in an economically rational fashion, Environmental Defense foresees a dramatically increased role for renewable energy and for energy efficiency. At the same time, since any complete transition away from fossil fuels is likely to take a very long time, we foresee a long-term need to deal with CO_2 emissions from coal-based facilities. The sooner we begin to deploy CCS technology on a large scale the better. We applaud you for working on measures to make this a reality.

The Chairman. Thank you very much.

Mr. Evans.

Statement of Ronald T. Evans, Senior Vice President, Reservoir Engineering, Denbury Resources, INC., Plano, TX

Mr. Evans. Thank you Chairman Bingaman and members of the committee for the opportunity to share our views on the policy aspects of carbon capture and storage for CCS. As Denbury Senior Vice President I oversee all reservoir engineering, land, acquisitions and purchases of anthropogenic CO_2. Denbury's primary focus is enhanced oil recovery utilizing CO_2.

We are currently the largest oil producer in the State of Mississippi and one of the largest injectors, if not the largest injector of CO_2 in terms of volume in the United States. Since 1999 we have produced over 15 million barrels of oil from CO_2 flooding from ten active EOR projects in Mississippi and Louisiana. We are currently participating in several demonstration projects and DOE's regional carbon sequestration partnership program. I will briefly address what we at Denbury believe are the most important policy aspects of carbon capture and storage: cost, taxation and the question of pipeline access and the legislation before the committee today.

Cost. Perhaps the single largest obstacle to developing CCS beyond a limited number of projects currently in operation is the significant cost involved with carbon capture and storage. The cost of capture stem from variations in the quantity and the quality of the CO_2 produced by hydrocarbon combustion, gasification or other industrial processes. The cost to purchase the compressors and the power to generate the compression necessary to pressure the gas significantly to enter the pipeline or sequestration, the lower the percentage of CO_2 in the stream of gases and the greater amount of impurities in the streams the greater the cost of capture.

In addition most technologies capture CO_2 at a lower pressure than the pressure required to enter a typical CO_2 pipeline or to inject into a deep saline reservoir or EOR project. The cost of the compressors and the power necessary to drive them are significant. One example approximately seven dollars and fifty cents per ton or just over one third of the estimated total cost of 20 dollars per ton for carbon capture and storage from the least expensive sources when transported only moderate distances.

The costs of transportation are also significant. Installation costs for CO_2 pipelines have increased dramatically in recent years. From about 30,000 dollars per inch mile for Denbury's free State pipeline to an estimated 100,000 dollars per inch mile for Denbury's proposed green pipeline due to rising steel prices, rising energy prices and construction costs doubling our effective CO_2 pipeline transportation rate. Without some means of reducing the cost of carbon capture and storage infrastructure, significantly development, will likely remain stagnant.

Senate bill 2144 directs the Secretary of Energy to study technical and financing issues related to the construction and operation of CO_2 pipelines. While further studies should prove useful, Congress can act now to address carbon capture and storage costs. Congress should amend section 7704, the tax code to clarify that section (d)(1)(E) covers man made as opposed to just naturally occurring CO_2.

A substantial portion of all the CO_2, natural gas, oil and product pipelines in the United States are owned and operated by publicly traded partnerships under section 7704 whose lower cost of capital lowers the cost of development and transportation of natural resources. Because of the current uncertainty in section 7704 much of the existing CO_2 pipeline capacity cannot be used and new capacity may not get built to transport anthropogenic CO_2. The Senate Finance committee approved a clarification last June, but Congress failed to include it in the Energy Independence and Security Act. We strongly urge members of this committee to work with their colleagues to pass this clarifying amendment.

Pipeline access. The natural gas, oil and product pipeline systems today consist of hundreds of thousands of miles of pipelines with significant interconnects between individual pipeline systems. There also exists a huge retail market or oil and natural gas with a large number of users. This situation stands at market contrast to CO_2 pipelines. In addition to CO_2 not being explosive, flammable or poisonous there currently exists no large interconnected system nor are there reasonable prospects for development of a retail market for CO_2 with a large number of users.

Only a limited number of regional CO_2 shippers and users exist. CO_2 pipeline systems are only a tiny fraction of the size of the oil and gas network. CO_2 pipelines should be given room to grow before FERC like regulation, including regulating access, is contemplated.

To conclude the U.S. economy will continue to require massive amounts of energy well into the future and thus this country needs to use all of its resources to produce the energy it requires given economic and environmental realities. EOR is already playing an important role in this regard and can do so by far greater scale with the right policies. EOR is the only currently active, actual on the ground method for CO_2 injection and sequestration.

While we agree that the additional research and studies proposed in Senate bills 2144 and Senate bill 2323 are worthwhile. We do not believe there's a need for comprehensive Federal regulation as section five of Senate bill 2323 proposes.

Congress should provide necessary incentives in mechanisms to foster the development of CCS allowing states to continue to oversee various aspects with which they already have significant experience. Thank you.

[The prepared statement of Mr. Evans follows:]

Prepared Statement of Ronald T. Evans, Senior Vice President, Reservoir Engineering, Denbury Resources, Plano, TX

Denbury Resources, Inc., (''Denbury'') appreciates this opportunity to share with Members of the Senate committee on Energy and Natural Resources its views on policy aspects of carbon capture, transportation, and sequestration (hereinafter collectively referred to as ''CCS''). As Senior Vice President, Reservoir Engineering for Denbury, I oversee all reservoir engineering, land functions and acquisition activities; am responsible for securing and contracting sources of anthropogenic CO_2; and coordinating our government relations. Denbury is currently the largest oil producer in the State of Mississippi and the one of the largest injectors of carbon dioxide (''CO_2'') in terms of volume in the United States. Denbury's primary focus is enhanced oil recovery (''EOR'') utilizing CO_2. At the present time we operate ten (10) active CO_2 enhanced oil projects, nine in the State of Mississippi and one in the State of Louisiana.

Denbury also owns the largest natural deposit of CO_2 east of the Mississippi River, called Jackson Dome in central Mississippi, which we extract and transport through approximately 350 miles of dedicated CO_2 pipelines for use in EOR. Denbury is also in the process of designing or constructing an additional 375 miles of CO_2 pipelines in order to expand our operations into additional fields throughout the Gulf Coast of the United States. Finally, the committee may be interested to know that Denbury is working with the federal Department of Energy and various research universities on several Phase II and Phase III demonstration projects in the Regional Carbon Sequestration Partnership Program. While our business model focuses primarily on the transportation and sequestration components of CCS, we also are very familiar with the capture component both in terms of (1) the compression demands of transportation and sequestration and (2) our enhanced oil operations, which recycle large volumes of CO_2 in order to recover additional volumes of oil. Given this background, Denbury is pleased to share its perspective on various policy aspects of CCS and the proposed legislation before the committee today.

A thorough understanding of both (1) the physical processes by which CO 2 is obtained, transported and injected for purposes of EOR and/or permanent storage, and (2) the economics that underlie existing and future EOR-related use of CO2 is essential to any consideration of potential policy issues. The significant and varying costs associated with CCS—whether in conjunction with EOR or not—are perhaps the single largest obstacle to developing CCS infrastructure beyond the limited, discrete projects currently in operation. From Denbury's perspective, it is critical that any contemplated state or federal regulation not increase these costs and impede private sector development of the CCS infrastructure necessary to meet the demands of our energy hungry and potentially carbon-constrained world. As explained in greater detail below, the current regulatory structure surrounding CO_2 consists of state and federal provisions that cover discrete aspects of CCS. For instance, the over 3,500 miles of dedicated CO_2 pipelines currently in use were constructed and are operating under rules and guidelines for safety issued by the Department of Transportation's Office of Pipeline Safety; with pipeline siting issues significantly impacted by state eminent domain laws; and with CO_2 injection wells permitted and approved by individual state government divisions or departments of Underground Injection Control, utilizing the standards and policies issued by the Environmental Protection Agency. While this system may appear patch-work and noncomprehensive, the current structure is entirely appropriate, as CCS is very much still in its infancy. This predominantly state-law-based system should suffice for many years to come. Thus, Denbury supports the recommendations of the Interstate Oil and Gas Compact Commission's 2005 Regulatory Framework for States. With few exceptions, such as funding research and further study of the issues involved as both bills propose, and given the current system of regulations and natural physical and economic constraints likely to exist for years to come, federal policymakers might best further national energy and carbon capture goals by deferring broad legislation or regulation while CCS is in this nascent phase.

I. Capture/Compression

In thinking about the policy aspects of CCS, it is useful to separate the various components of CCS and to identify what issues within each merit particular attention, distinguishing between EOR-related CCS and CCS in saline or other formations where appropriate. The starting point for any type of CCS is to capture the CO_2. Denbury currently obtains all of its CO_2 from its natural deposit at Jackson Dome. Certain existing and some evolving technologies allow CO_2 emitted from various manufacturing processes to be captured. The combustion or gasification of hydrocarbon-based fuels such as coal, petcoke or other hydrocarbons produces particularly large volumes of CO_2 at varying levels of quality and purity. As new capture-inclusive projects are constructed, Denbury plans to acquire thousands of metric tons of CO_2 each day for use in EOR.

Aside from the threshold questions of how to properly classify CO_2 and whether and to what extent to restrict emissions, from Denbury's perspective, the capture of CO_2 presents no policy issue. Rather, the capture component presents a significant economic issue: First, capture technology is expensive. The byproduct of hydrocarbon combustion or gasification is a stream of gases and other impurities that contains various quantities of CO_2. In order for CO_2 to be usable in EOR it must be injected in a relatively pure form. Similarly, CO_2 injected into deep saline reservoirs must be in a relatively pure form to maximize the storage space available to be filled with CO_2. Thus, a significant component of the capture cost is the cost to separate and purify the CO_2 to be injected. The lower the percentage of CO_2 in the stream of gases and the greater the amount of impurities in the stream the greater the cost of capture. Second, most technologies capture the CO_2 at a lower pressure than is required to either enter a typical CO_2 pipeline or to inject into a deep saline reservoir or EOR project. The costs of the compressors and the power necessary to drive them are significant—approximately $7.50/ton of the estimated $20/ton total cost[1] for CO_2 that is transported moderate distances. Therefore, the compression costs associated with CO_2 capture are slightly more than one-third (33%) of the total CCS cost for the least expensive sources of anthropogenic (man-made) CO_2. Additional compression costs are incurred to maintain pressure in pipelines and again when CO_2 is pressured up to sufficient level for EOR reservoir injection. In sum, without some means of reducing the cost of captured anthropogenic CO_2 significantly, infrastructure development will likely remain stagnant.

To address this issue, last year the Finance committee approved a tax credit forthe capture and sequestration of CO_2 of $10.00/ton in connection with EOR and $20/ton for non-EOR projects for up to 75,000,000 tons sequestered. From Denbury's perspective, this would be sufficient to incentivize construction of additional pipe-lines from emission sites to geologic sequestration sites in connection with EOR activities. Unfortunately, this provision was not included in the energy legislation ultimately signed into law in December. We hope that Congress will address the issueof CCS costs in 2008, especially those associated with capture and compression, and note that proposed projects from gasification through to sequestration have the potential to create hundreds and perhaps thousands of jobs across the country. On this point, S. 2144 directs the Secretary of Energy to study technical and financing issues related to the construction and operation of CO_2 pipelines and sequestration facilities. While this will be helpful to policymakers, the legislation should also direct the Secretary to consider these same issues in relation to CO_2 capture, separation, purification and compression.

II. TRANSPORTATION

The most economical way to transport CO_2 is through pipelines at pressures in excess of 1100 psi so that the CO_2 is transported as a supercritical fluid (dense phase). At pressures in excess of 1100 psi and temperatures common for CO_2 pipelines, CO_2 is a supercritical fluid which means that the CO_2 has properties of both a liquid and a gas. Larger volumes of CO_2 can be transported through CO_2 pipelines in this dense phase than can be transported as a gas. Given the pressure requirements to maintain CO_2 in the dense phase, CO_2 pipelines are generally operated at pressures greater than 2,000 psi. This pressure is well in excess of the average operating pressure of a natural gas pipeline, though the material used to manufacture both types is the same.

A. Safety

CO_2 is not as dangerous to transport as some other gases, such as hydrogen and natural gas because it is not explosive, flammable or poisonous. The primary safety issue with transporting CO_2 is asphyxiation caused by a leak in a pipeline. Although there have been a few accidents, releases and leaks reported, none of the dozen leaks that occurred from 1986 to 2006 resulted in significant injury. The characteristics of anthropogenic CO_2 and natural CO_2 are essentially the same. Thus, whether natural CO_2 or anthropogenic CO_2 is being transported in a CO_2 pipeline for the purposes of being delivered to an enhanced oil recovery project or being delivered to a deep saline reservoir sequestration project is irrelevant to the safe construction and operation of a CO_2 pipeline. At the present time there exist over 3,500 miles of dedicated CO_2 pipelines, most of which have been transporting CO_2 for over 20 years—and some for over 30 years—with an excellent safety record. We do not see any evidence to suggest that the current regulatory framework that oversees construction and operation of CO_2 pipelines should be modified. To the extent that consideration of safe handling, transportation, and sequestration issues by the Department of Energy, as S. 2144 directs, will address any lingering misconceptions about the relative safety of dense phase CO_2, it will facilitate public understanding and acceptance of CO_2 pipelines and sequestration projects.

B. Siting

At the present time federal eminent domain authority does not extend to CO_2 pipelines. Several states have provided eminent domain authority to CO_2 pipeline owners to assist in getting CO_2 pipelines constructed. While this is helpful in constructing *intrastate* pipelines, individual state eminent domain powers may not extend to interstate pipelines that are just traversing through a state with no origin or terminus there. For this reason and due to the long distances across state lines that separate potential CO_2 emission capture sites from potential EOR locations, federal eminent domain authority may ultimately be required to develop a nationwide CO_2 pipeline infrastructure. In addition, some mechanism may be necessary to address the siting of pipelines and CCS generally on federal lands. S. 2144 directs the Secretary of Energy to study CO_2 pipeline siting issues, which should facilitate a thoughtful approach by policymakers.

C. Rates

Any contemplation of federal regulation of CO_2 transportation rates and pipelines similar to the regulations that currently exist for natural gas, oil or products pipelines is premature, as there is no interconnected system of CO_2 pipelines to which to apply any such regulation, nor prospects for development of one for many years, nor reasonable prospects for development of a ‘‘retail’’ market for CO_2 with large numbers of ‘‘users’’ of the CO_2. At the present time there are very limited areas with existing CO_2 pipelines and limited industrial CO_2 emissions being captured (North Dakota Gasification). The vast majority of the existing CO_2 pipelines are transporting natural CO_2 from natural underground CO_2 production sources that are owned and operated by the CO_2 pipeline owner—generally for use in enhanced recovery projects also owned and operated by the CO_2 pipeline owner. In cases where the owner of the CO_2 pipeline has CO_2 production volumes in excess of its own EOR requirements, the excess CO_2 volumes are sold to EOR operators in other projects or to industrial gas suppliers. This limited number of regional CO_2 shippers and consumers stands in marked contrast to the numerous and geographically widespread producers and consumers of oil and natural gas products.

It would be a substantial mischaracterization to suggest that the U.S. has an integrated CO_2 pipeline system similar to the fully integrated natural gas, oil or hydrocarbon products pipeline systems which have their transportation rates regulated by the Federal Energy Regulatory Commission (''FERC''). The natural gas, oil and product pipeline systems today consist of hundreds of thousands of miles of pipelines with significant interconnects between individual pipeline systems to accommodate the transfer of natural gas, oil or products from one pipeline system to the other. In contrast, existing CO_2 pipeline systems are a tiny fraction of that size (3500 miles) and are not interconnected. (see Attachment No. 1) Several pipelines delivering CO_2 for enhanced oil recovery in the Permian basin of west Texas are interconnected at Denver City, where CO_2 can be transferred from one pipeline to another. The other CO_2 pipeline systems in Wyoming, North Dakota, Oklahoma, and Mississippi are not connected to the Permian basin pipeline system or to each other. Thus, today no national CO_2 pipeline system exists and no federal regulation to ensure access is necessary.

Natural gas, oil and hydrocarbon products pipelines were constructed in a similar manner to today's CO_2 pipeline systems. Individual pipeline systems were developed to transport natural gas, oil or products from production sites to consumption sites in their infancy. Only after a significant period of time, were these individual systems eventually interconnected to allow the transfer from one pipeline system to the other. Although the Federal Power Commission and eventually the FERC was granted jurisdiction over the transportation rates for natural gas, oil and hydro-carbon products, the combination of regulating rates and requiring open access has only existed since 1985. Several decades passed between the time that individual pipelines were constructed and eventually interconnected to create an integrated intrastate pipeline system. CO_2 pipelines should also be given room to grow before FERC-like regulation is contemplated.

D. Costs

The construction and installation of CO_2 pipelines is a capital intensive effort, the costs of which have increased in recent years for a variety of reasons, including rising steel prices, construction costs and energy prices. By way of example, Denbury's 93 mile, 20 inch Freestate pipeline (see Attachment No. 2) completed in 2006 cost approximately \$30,000 per inchmile, resulting in an effective transportation rate of approximately \$3.50/ton at full capacity. The initial

37 mile segment of Denbury's 24 inch Delta pipeline was completed in 2007 at a cost of approximately \$55,000 per inch-mile. We estimate that our planned 314 mile, 24 inch Green Pipeline that will run from Donaldsonville, Louisiana to Hastings field in southeast Texas will cost approximately, \$100,000 per inch-mile resulting in an effective transportation rate of approximately \$7/ton at full capacity. While the length (pumping stations to maintain adequate pressure add an additional \$1 to \$2 per ton to transportation costs), route obstacles and type of terrain all added to the estimated cost of the Green pipeline, the fact remains that such endeavors, even under the best of circumstances are extremely costly and take years of careful planning. As stated above, S. 2144 directs the Secretary of Energy to study technical and financing issues related to the construction and operation of CO_2 pipelines. Such information should prove useful to policymakers seeking to understand the significant costs involved in developing the infrastructure of CCS. Also, any study of CO_2 pipeline financing issues will undoubtedly encounter the tax code impediment discussed in the next section.

E. Taxation

Today, a substantial portion of all CO_2, natural gas, oil and products pipelines in the U.S. are owned and operated by companies that are organized as Publicly Traded Partnerships commonly referred to as Master Limited Partnerships (''MLPs''), which through their lower cost of capital have been an important financing source for building these assets. Section 7704 of the tax code permits MLPs to be taxed so that income and tax liabilities are passed through to the partners, even though the MLPs are large public entities, provided 90 percent or more of the MLP's gross income is derived from certain qualifying activities. These activities include exploration, development, processing and transportation of natural resources, including pipelines transporting gas, oil, or products thereof (see Sec. 7704(d)(1)(E)). While this provision covers the processing and pipelining of ''natural'' CO_2, it is unclear whether it covers anthropogenic CO_2. Because of this uncertainty, much of the existing CO_2 pipeline capacity (that owned by MLPs) cannot currently be used to transport anthropogenic CO_2 from emissions sites—at least not without significantly higher tax costs than other pipeline assets in the industry.

Last year, as part of its energy tax package, the Senate Finance committee adopted a modification to include industrial source CO_2 in the definition of qualifying income (see Sec. 817 of the Energy Enhancement and Investment Act of 2007, June 19, 2007). However, Congress ultimately failed to include that package of provisions in the Energy Independence and Security Act of 2007 (P.L. 110-140). Without this modification of the tax code, a substantial 6 portion of the pipeline industry will most likely not contribute capital to the construction of the CO_2 pipeline infrastructure necessary to facilitate CCS through transportation of anthropogenic CO_2. We strongly urge Members of the Energy and Natural Resources committee to work with their colleagues on the Finance committee and the House Ways and Means committee to accomplish this important clarification.

III. INJECTION / SEQUESTRATION

Enhanced oil recovery utilizing CO_2 requires multiple injection wells throughout a unitized field or reservoir. CO_2 injection wells are permitted and approved by each State's division or department of Underground Injection Control utilizing the standards and policies issued by the EPA. CO_2 injection wells utilized in tertiary oil recovery (a.k.a. EOR) are permitted and approved as Class II Injection wells. Such wells have been in existence for over 30 years. The CO_2 sequestration commercial demonstration projects proposed in S. 2323 and enacted in the Energy Independence and Security Act of 2007 should yield additional helpful data on the ability of EOR and saline reservoirs to sequester CO_2.

In 2005, the Interstate Oil and Gas Compact Commission (''IOGCC'') issued its recommendations concerning CO_2 injection wells in EOR and non-EOR applications. The IOGCC has recommended that future CO_2 regulation should build upon the primarily state-based regulatory framework already in place, due to states' decades of experience with CO_2 EOR, natural gas storage, and acid gas injection. We concur with their recommendation that for future CO_2 injections in EOR projects, the existing regulatory framework should not be modified. The IOGCC recommended that for non-EOR CO_2 injections, additional regulatory requirements may need to be considered since these types of applications may not have a defined period of injection as does EOR. We also concur with the IOGCC recommendation that CO_2 injection wells for non-EOR applications should be permitted and approved as a sub-class of Class II injection wells or a new classification but not permitted as Class I or V injection wells.

Generally, every CO_2 well drilled is required by state regulations to set and cement a surface casing string below the Underground Source of Drinking Water (USDW) depth to protect the fresh water and ground water intervals. Cement is required to be circulated back to the surface to insure that all potential zones above the USDW depth that contain freshwater are protected. Only after setting the surface casing are wells drilled to the depth required to produce oil and gas or to inject CO_2. Once the well reaches total depth an additional casing string is cemented in the well to provide additional protection to the freshwater intervals and to produce or inject through. We believe existing laws and regulations provide sufficient protection of the fresh water and ground water reservoirs from the injection of CO_2 in EOR operations or, for that matter, in deep saline reservoirs.

The potential for significant migration or leakage from an EOR operation is extremely remote due to the geological nature of oil and gas reservoirs and the existing mechanism that has trapped the oil or gas. At the present time oil and gas operators are required under their mineral leases and state regulations to properly plug and abandon wellbores within a reasonable period after oil and gas operations cease. Responsibility for re-plugging an improperly plugged well remains with the oil and gas operator for an extremely long period of time and, in practice, remains as long as the oil and gas operator is in existence. Such responsibility should be essentially the same for deep saline reservoir injection. However, the detailed geologic and engineering information required by states for EOR projects does not exist for saline reservoirs. Thus, information about deep saline reservoirs will have to be developed, taking into account that CO_2, being less dense than saline water, will segregate dueto gravitational forces and migrate to the highest subsurface position in the reservoir. As noted above, S. 2323 proposes, and the Energy Independence and Security Act of 2007 provided for, commercial demonstration projects, as well as a national CO_2 storage capacity assessment. These undertakings should yield important data currently lacking on saline reservoirs.

IV. CONCLUSION

The U.S. economy will continue to require massive amounts of energy well into the future and thus the country needs to use all of its resources to produce the energy it requires given economic and environmental realities. EOR is already playing an important role in this regard—taking a waste product and using it to increasedomestic energy production—and can do so on a far greater scale, with little action required by federal policymakers. The most important step Congress can take at present is to amend Section 7704(d)(1)(E) of the tax code to make clear that anthropogenic CO_2 is included.

The two bills being considered by the committee today, S. 2144 and S. 2323, are clearly intended to provide meaningful vehicles to better understand the issues central to CCS and we commend the committee for focusing on them. While we agree that additional research and further study are worthwhile—as both bills propose— we do not believe there is a need for comprehensive federal regulation, as Section 5 of S. 2323 proposes. Of course, there are areas where federal oversight will likely be necessary, such as management of CO_2 on and under federal lands. For the most part, however, Congress should simply provide necessary incentives and mechanisms to foster the development of CCS, allowing states to continue to oversee various aspects with which they already have significant experience.

The Chairman. Thank you very much. Thank you all for your excellent testimony.

Let me start with you, Mr. Bengal. I think you make a point in your testimony which is fairly key to our consideration of this whole subject and that is that you say States and provinces are likely to continue to regard CO_2 geologic storage reservoirs as a valuable resource that should be managed using resource management frameworks therefore voiding the treatment of CO_2 storage as a waste disposal. I gather that what you have put together, your task force as you see it, deals with this not as a pollutant but rather as a resource that should be managed in that way. Could you clarify that for me?

Mr. Bengal. The CO_2 in and of itself may be not the resource as much as the pore space you would be putting it into. Primarily there's only so many places that would be good for CO_2 storage. Not every State has good geologic sites and there's some States that you should not store CO_2 in because of the nature of the geology. It's just not safe and sound.

So there will be a competition for that pore space for those places where it is good that would be an economic benefit to a State who has that pore space to effectively manage that pore space. Ensure that No. 1, the maximum amount of CO_2 is put into that storage area so it's not wasted, the space is not wasted, to keep other entities from encroaching upon that area so it can be set aside just for a particular project. There's a question about drilling through that site. If you had an entire site permitted and set aside for any particular project the State would then ensure that there's no other penetration to the well bores to that site.

So what we're really talking about is we're looking at the management of the pore space where you put the CO_2 as basically the resource. The CO_2 placed in that resource management frameworks deal with that because we deal with natural gas storage maximizing pore space to store the natural gas. With oil and gas you maximize recovery by managing the pore space much the same way here.

In a waste framework you're just looking at a place to inject something to get rid of it. You're not managing effectively what you're using and where it's going. So that's the way we want to look at it. It's more regulatory framework issue as opposed to the CO_2 itself.

The Chairman. Let me ask, with my very limited knowledge of this subject at this point, it seems to me that a big problem is that much of the production of CO_2 from power plants is not going to be particularly near the storage locations that we're going to try to store or sequester this CO_2 in. So we're going to be talking about quite a few pipelines that are transferring this CO_2 across several states. So there's going to be a Federal responsibility once you get an interstate pipeline.

What do you see? I know you have a thing here saying that broader, overarching Federal regulation that's like that contemplated in 2323 is not appropriate in your view. How do you see the Federal Government regulating that transportation if I'm right that the significant amount of interstate transportation is going to be required?

Mr. Bengal. What we're referring to basically is the storage site itself and not the transportation system. That would be a Federal role. The pipelines and the infrastructure for that would remain a DOT role as it is or a FERC role.

We're basically talking about the regulation of the site, the licensing of the site, the long term storage and things like that is what we're referring to and the Federal regulations that would deal with that. With respect to the location and the cost of CO_2 from existing power plants, you're absolutely correct. The cost of right now to retrofit an existing power plant to concentrate the CO_2 from that emission stack and then transport it somewhere, a distance for storage which is many years off in the future for sure.

What we're looking at is initially, I think the first projects would be basically a plant built for that purpose at a storage site where you have minimal transportation. Those are the kinds of projects we need to get going on first and right away as opposed to planning for this massive retrofitting of all existing power plants and a massive pipeline system which we don't need to do first. We need to get some major projects going right now. The technology exists. The regulatory frameworks are in place right now to do that.

The Chairman. Senator Corker.

Senator Corker. Yes, sir, Mr. Chairman. Thank you for your testimony, all of you. Mr. Anderson, you talked about how coal is going to be moving into a new era. I can't help but think, based on all the complications that center around either the cost of transport that Mr. Evans talked about and just the geographic differences that exist between where carbon is produced and where it's going to be stored, that really, unless there's huge allowances that are laid out for coal on the front end, that basically coal is going to go through a really, really tough period of time beginning in 2012 if a bill does pass regarding carbon cap and trade which I'm not saying is a plus or minus. I'm just making an observation. Do you have any comment on that? It just seems to me that the price of coal, the cost of coal produced electricity is going to skyrocket in the beginning as some of these other more complicated things are worked out. I wonder if you have any comment in that regard.

Mr. Anderson. The studies that I'm familiar with project a very large role for coal based CCS in the future under regimes like this. The International Energy Agency for example has estimated that by 2050 that CCS could rank second only to energy efficiency as a contributor to solving global warming. There are other estimates that project a 15 percent to 55 percent contribution.

So while I think that business as usual is over for coal, I think coal has a very bright future.

Senator Corker. That it has a bright future just at a different cost structure?

Mr. Anderson. Yes, sir.

Senator Corker. Mr. Evans do you have any comments in that regard?

Mr. Evans. The comments I have, Senator, would be that the actual cost of transportation when you look at it although it's one third of the cheapest sources, it becomes a much, much lower component of say from an existing coal fired power plant. It may be as low at 10 percent of that cost. So really the cost on conventional coal today is primarily in the capture side not the transportation.

As we develop sequestration in general with EOR the oil and gas companies can build pipelines to capture the CO_2 and transport to their oil fields. We can cover that cost of the transportation and the sequestration side. It's how much of the capture cost in addition to that are we able to cover.

Senator Corker. What kind of commercial market for carbon other than for use in enhanced oil recovery do we see 15, 20 years out? What part of the carbon that will be produced can actually be used for other commercial uses other than enhanced oil recovery?

Mr. Evans. If I do a comparison in Mississippi we produce almost 550 million cubic feet a day of CO_2. About 80 million cubic feet of that goes into industrial uses to make dry ice, freeze chickens, industrial uses of CO_2 so there you're looking at around 20 percent. That market has only been growing about two to 3 percent a year. So I don't know that there's going to be without significant discoveries of other uses for CO_2 much use of it other than EOR or permanent sequestration.

Senator Corker. Mr. Chairman, thank you.

The Chairman. Thank you.

Senator Lincoln.

Senator Lincoln. Thank you, Mr. Chairman. Again welcome gentlemen. We appreciate your expertise in working with us.

Mr. Bengal you mentioned some states have begun the process of beginning a legal and regulatory structure for carbon storage. What are some of those or who are some of those States?

Mr. Bengal. California, New Mexico, Wyoming, North Dakota, Texas and several other States.

Senator Lincoln. So there are a few.

Mr. Bengal. Yes, they're working on legislation as well as rules to move CO_2 storage along in their States.

Senator Lincoln. What was the other or whoever was on your task force? Was there other expertise there?

Mr. Bengal. Yes. The task force consisted of actually of two task forces through phase one and phase two. But it was State oil and gas regulators from various oil and gas States, representatives from the various DOE partnerships, the Association of Professional or state geologists, State geologic surveys, representatives from DOE, EPA, BLM and Environmental Defense was an observer during the process of our rules development.

Senator Lincoln. So you had quite a wide——

Mr. Bengal. Yes. There were industry experts as well.

Senator Lincoln. I don't know that you've gotten into this yet today and maybe I'm being redundant. I hope not. But you might explain to me the issue regarding storage rights.

Mr. Bengal. Much like natural gas if you're going to store something underground like you do natural gas the pore space is owned by someone. It's mineral right. It's a property right and you can't just use that without the property rights authority to do so.

That has to be acquired and that's generally, probably akin to the natural gas storage industry where it belongs to the surface owner. So in order to store these amounts of CO_2 we're talking about even though you may have a very large area an operator will have to acquire the right to store it in that area. There was a question before where's the plume going to go? How are you going to manage the plume for future liability?

You're going to know that before you start because that operator is going to have to own and control the entire area where that plume will go. That's a very large undertaking to do to acquire those rights. So that will be worked out prior to injection what the ultimate static disposition of that plume will be because it will be owned and controlled by owning those property rights.

Senator Lincoln. That seems like that would be quite the lengthy process. I know just with the Fayetteville shale drilling that is going on in Arkansas the mineral rights and property rights and how they've gone in there has taken quite a bit of time.

Mr. Bengal. It will be in a natural gas storage setting. States do that through eminent domain condemnation proceedings. You get a certain percentage of the site locked up and then you would go to the State and condemn the rest to move that project forward.

Senator Lincoln. In terms of the—maybe you can help explain a little better too, more detail, how you see the state framework working with the EPA framework? Is that kind of like a MOA or how do they do that?

Mr. Bengal. Right now the UIC program, the Underground Injection Program, most states have the authority through privacy from the U.S. EPA to administer that program at the State level for the U.S. EPA in each state.

Senator Lincoln. So they've already got that?

Mr. Bengal. They've already got that. In a few States, a direct implementation State, the EPA does on its own. But for those States that do have privacy whatever EPA changes the regulations to be they're automatically incorporated into the State functions that exist and are ongoing now.

Senator Lincoln. Is there something special about Federal land. Is there something that—does there need to be regulations of CO_2 storage specifically on Federal lands?

Mr. Bengal. They probably will have to develop a similar type of framework for regulation that we would have on private lands within the States. The State would probably be involved in that. I don't see why it would have to be much different than what the States would do on private lands. The one benefit of Federal lands is you have one mineral owner or one surface owner for the entire project. You could actually site projects very succinctly and not have to deal with the property issue at all.

Senator Lincoln. Last just about emissions trading or accreditation of storage projects for the purposes of securing the carbon credits. Does your proposed infrastructure cover that as well?

Mr. Bengal. No. We do not do that. We dealt with just the technology and the legal framework for the storage itself. That's probably going to be worked out at the marketplace or be federally— under the cap and trade system and you'll have to figure that out. We will fit in whatever it is.

Senator Lincoln. We're always grateful for any suggestions or models you might have already come up with.

Mr. Bengal. Just don't do more than you have to.

Senator Lincoln. Ok. Thank you, Mr. Chairman.

The Chairman. All right. I think that's the end of our questions, and thank this panel very much. This was very useful testimony and we think it is a useful hearing. Thank you.

[Whereupon, at 4:25 p.m. the hearing was adjourned.]

APPENDIX I :RESPONSES TO ADDITIONAL QUESTIONS

Responses of Scott Anderson to Questions From Senator Bingaman

Question 1

Many people have called for ''permanent'' storage in introduced legislation (e.g. all the introduced climate bills), but do not go so far as to define permanence. Can any one of you elaborate on a clear definition of ''permanent CO_2 storage'' as it relates to geologic storage?

Answer. We suggest using retention of at least 99% for at least 1000 years as the standard for ''permanence.'' The Intergovernmental Panel on Climate Change has indicated that it is feasible to meet such a standard in well-selected, well-managed sites.

Question 2

What is the appropriate amount of time that CO_2 should be stored in the subsurface, as means of mitigating CO_2 emissions? Will storage times of that magnitude be burdensome on CCS projects?

Answer. We are hopeful that a 1,000 year standard as suggested above will assure that any material leakage that takes place will occur well beyond the era of fossil fuels and therefore at a time when climate change due to emissions of greenhouse gases such as CO_2 is no longer an issue. We do not expect such a standard to be burdensome and in fact we expect that well-executed projects will have minimal leakage for even longer times. Studies at the Weyburn project, for example, indicate a 95% probability that 98.7 to 99.5% of the sequestered CO_2 will remain in the geosphere after 5,000 years and that there will be even less leakage as time goes on. The longer CO_2 is contained underground the more likely it is to stay contained due to mineralization, dissolution in formation water, and residual trapping in pore spaces. See Whittaker, S., White, D., Law, D, and Chlaturnyk, R., IEA GHG Weyburn CO_2 Monitoring & Storage Project Summary Report, Petroleum Technology Research Center, Regina, 273 (2004); Damen, K., Faaij, A., and Turkenburg, W., Health, Safety and Environmental Risks of Underground CO_2 Storage—Overview of Mechanisms and Current Knowledge (Springer 2006).

Question 3

Is there any amount of leakage that is acceptable? The IPCC suggests that storage should be on the order of 1000 years in a geologic formation, with less than 1% leakage of the volume of CO_2 that is injected over the life of the storage project. Is this a reasonable expectation? How can we enforce such a requirement?

Answer. Some have calculated that leakage rates as low as .01% per year, implying 90% retention over 1000 years, might be acceptable from the perspective of climate change policy. Hepple, R. and Benson, S., Implications of Surface Seepage on the Effectiveness of Geologic Storage of Carbon Dioxide as a Climate Change Mitigation Strategy, in Gale, J. and Kaya, Y. (eds), Sixth International Conference on Greenhouse Gas Control Technologies, vol. I, 261–266 (2003). We believe that the 99% for 1000 years standard is feasible and provides a better margin of safety.

In order to enforce such a requirement, it will be necessary to: (1) require that sequestration projects take place only at well-characterized, properly selected sites; (2) require project developers to define the containment system and explain why it is reasonable to expect the system to contain the appropriate amount of CO_2 for the appropriate time frame; (3) require the developer to model and project the fate of the CO_2 in the containment system; (4) require monitoring to confirm or modify the definition of the containment system and to confirm or modify the projections regarding the fate of CO_2 in the system; (5) require that operations be modified if monitoring indicates that a risk of unacceptable leakage is developing; and (6) require remediation where problems develop that cannot be adequately resolvedthrough modification of operations.

Question 4

In your written testimony, you support the coordinated efforts of the EPA, DOE, and DOI that are specified in Senator Kerry's bill (S.2323). Would there be any additional agencies, organizations, or individuals who you feel should be involved in the interagency task force? (e.g. in the case of ultimate liability of storage sites the Department of Justice may need to be consulted, DOT & FERC for pipe-lines issues, etc.)

Answer. Yes, the Department of Justice, DOT and FERC could all have valuable roles to play.

Question 5

You also state in your testimony a need to specify ''permanent'' geologic storage. In your opinion and that of Environmental Defense, how would you define permanent storage?

Answer. Please see answer to question 1 above.

Responses of Scott Anderson to Questions From Senator Dorgan

Question 1

I also think it's important that the general public have an understanding of how vastly important an issue this will be to our energy future. In your judgment, what will it take for the general population to better understand and support the approaches associated with carbon capture and storage?

Answer. Education is important in order to achieve public understanding and support. A sound, rigorous regulatory framework is also essential. In addition, the public is more likely to support CCS if it views the technology as only one tool among many for combating climate change, as opposed to viewing CCS as being in competition with renewables and energy efficiency.

Question 2

Coal and oil & gas are two different sectors that traditionally have little history of working together. Today, through enhanced oil recovery opportunities, we are seeing these partnerships beginning to take shape. What kind of new relationships and partnerships will have to be established in order to achieve larger scale carbon capture and storage projects? How can we build on these and other public and private sector relationships to expand this into an industry (regional or national)?

Answer. Putting a market value on carbon sequestration by passing cap and trade legislation is the single most important thing Congress can do in order to encourage these relationships. Once this is done, the market will be able to become the primary driver for answering questions about the types of relationships and partnerships that ought to arise.

Response of Scott Anderson to Question From Senator Menendez

Question 1

Mr. Anderson, I agree with your testimony that carbon capture and sequestration technology needs to be advanced and I agree with you that a cap and trade bill would be the best way to make sure this technology is developed quickly. But I also do not want us to lose focus on the fact that we need to eventually transform our economy from one based on fossil fuels to one based on renewables. I am curious about your views on what the balance between funding for CCS and renewable should be. For instance, the Administration is asking for more funding for research on the technology than they are asking for in solar research. Is this the correct balance in your opinion?

Answer. We support the proposed level of CCS funding because it seems proportionate to current RD&D needs. This funding level should not be permanent, however. Once these investments have been made, the investment incentive in the future will need to come from the price signal induced by cap and trade legislation. Publicly-funded research needs for renewable energy are likely to remain large for a longer period of time, and accordingly we hope that the balance between CCS investments and investments in solar and other renewable energy will improve as time goes on.

Responses of Benjamin H. Grumbles to Questions from Senator Bingaman

Question 1

Many people have called for ‘‘permanent’’ storage in introduced legislation (e.g. all the introduced climate bills), but do not go so far as to define permanence. Can any one of you elaborate on a clear definition of ‘‘permanent CO_2 storage’’ as it relates to geologic storage?

Answer. There is currently not a precise definition of ‘‘permanent CO_2 storage.’’ Effectiveness of geologic storage is contingent on CO_2 remaining stored underground for a long period of time. A desirable timeframe for geologic storage of CO_2 is on the order of thousands of years or longer. However, effectively sequestering CO_2 for even a few hundred years could provide valuable flexibility in reducing CO_2 emissions and contribute to reducing the costs of mitigating climate change.

Accumulation of CO_2 in natural geologic formations has been underway as a natural process in the earth’s upper crust for hundreds of millions of years. In most proposed carbon capture and storage (CCS) projects, the goal is to remove CO_2 from the atmosphere, and store it in the subsurface for significant periods. Thus, ‘‘permanent’’ means some long period of time that provides a reasonable assurance that the majority of the CO_2 will stay in place over a number of years (100’s to 1,000’s). Eventually, an increasing portion of CO_2 stored in the subsurface will be trapped through processes such as formation of minerals, and hydrodynamics with the result that this portion of the CO_2 would be sequestered at a geologic time scale of millions of years. As we gain knowledge from geologic storage projects, a more precise understanding of ‘‘permanent CO_2 storage’’ should emerge.

For the purposes of EPA’s proposed rulemaking under the Underground Injection Control (UIC) program, we have not defined this term. However, CO_2 will be stored for long periods of time (e.g., centuries) and EPA’s regulations will ensure that CO_2 storage is conducted in a manner that does not endanger underground sources of drinking water.

Question 2

What is the appropriate amount of time that CO_2 should be stored in the subsurface, as means of mitigating CO_2 emissions? Will storage times of that magnitude be burdensome on CCS projects?

Answer. DOE and others are conducting pilot scale geologic sequestration projects to help better understand these questions and others. CO_2 storage in geologic formations can mirror the timescale of oil and gas deposits in formations containing naturally occurring carbon dioxide gas. These formations have held these fluids for millions of years. A desirable timeframe for geologic storage of CO_2 is on the order of thousands of years or longer.

Demonstrating storage over these timeframes should not be overly burdensome. For well-selected, designed, constructed and managed geologic storage sites, the vast majority of CO_2 will gradually be immobilized by various trapping mechanisms and,in that case, could be retained for up to millions of years. Because of these mechanisms, storage could become more secure over longer timeframes (IPCC 2005).

Question 3

Is there any amount of leakage that is acceptable? The IPCC suggests that storage should be on the order of 1000 years in a geologic formation, with less than 1 % leakage of the volume of CO_2 that is injected over the life of the storage project. Is this a reasonable expectation? How can we enforce such a requirement?

Answer. EPA is currently deliberating on several of these issues as the Agency works to develop a regulatory proposal under the UIC program. EPA anticipates that some CCS projects will exhibit a certain amount of leakage within the sub-surface, however, safeguards such as leak detection and monitoring will protect against leakage that endangers underground sources of drinking water. The Agency is considering requiring monitoring for leakage to the atmosphere as it is an indicator of potential leakage to or endangerment of underground sources of drinking water.

Question 4

In the context of the national regulations that you are developing for carbon storage, have/will you make a clear definition of the appropriate storage time for CO_2 in the subsurface? Will you be proposing acceptable leakage rates and be mandating a minimum storage time? If so, how will you enforce this regulation?

Answer. EPA is currently deliberating on these issues as it works to develop a proposed regulation.

Question 5

In the case of other ‘‘pollutant’’ storage, the EPA has mandated 10,000 years residence time for the pollutant (or waste product) in the geologic subsurface. Do you anticipate proposing this sort of mandatory time restriction? In the case of Yucca Mountain this has proved to be very burdensome in the permitting/regulatory process.

Answer. EPA is in the middle of a deliberative process as it develops its proposed regulations and is considering this, among many other issues.

Question 6

According to a study conducted by Argonne labs, there is a perception from state EPA employees that they are currently inadequately trained, underfunded, and understaffed for handling the existing UIC program. After you complete the new regulations that you have described in your written testimony, do you anticipate that the state and regional EPA offices will have enough manpower, fiscal resources, training, and expertise to effectively implement these new rules? Should new staffing be required? How will this be funded?

Answer. While CO_2 storage raises new technical considerations, EPA is committed to continuing to support regional and state regulators for the purposes of implementing the UIC program. Currently, EPA provides nearly $11 million annually to assist primacy states and EPA regions where states do not have primacy with UIC program implementation. EPA directly implements programs in 10 states and shares responsibility in 7. The FY 2009 President's Budget requests $10.9 million for this work.

We would note that the Clean Air Act Advisory committee's (CAAAC) Advanced Coal Technology Work Group, which recently issued its final report, developed a rec-ommendation regarding this topic. Specifically, the Work Group recommended that EPA, working with other agencies, ''sponsor education and training programs for regulators and other officials involved in the permitting and monitoring of carbon capture and sequestration projects'' (more information is available at http://www.epa.gov/air/caaac/coaltech.html). EPA is currently preparing a more indepth evaluation of this issue in order to respond to CAAAC. The President's Budget requests a total of $3.9 million for UIC regulatory work in FY 2009, which the Agency could partially target to help address the CAAAC recommendation.

Responses of Benjamin H. Grumbles to Question from Senator Dorgan

Question 1

It seems to me that we need to much more quickly begin establishing and defining the ''rules of the road'' when it comes to carbon management. As we begin to unlock the opportunities for capturing, moving and storing larger amounts of CO_2, it is fair to say that the federal government will likely play a greater role. It will be better if we begin to better define appropriate roles for local, state and federal government. What are the most critical near-term issues that your agency can address so that developers can begin demonstrating CCS projects?

Answer. EPA understands the importance of clearly defining roles and responsibilities. Under the Safe Drinking Water Act, EPA develops minimum requirements for state and tribal Underground Injection Control (UIC) programs. Primacy states may develop their own regulations for injection wells in their state. These requirements must be at least as stringent as the federal requirements (and may be more stringent).

EPA has been working closely with DOE over the past several years as they implemented their CCS research and development program. The Agency recognized the critical near-term issue associated with facilitating UIC permits for demonstration projects to ensure that projects are carried out in a manner that does not endanger underground sources of drinking water. To address this, in March 2007, EPA issued guidance on permitting experimental projects as Class V injection wells. EPA plans to propose regulations in the summer of 2008 to ensure consistency in permitting fullscale CO_2 geologic sequestration projects. Final regulations would be issued by 2011. As with our other regulations, when EPA publishes new federal regulations with specific criteria and standards for constructing, operating, and closing CO_2 wells, a primacy state would need to adopt these standards and classes of wells and seek approval from EPA. In the interim, states will be able to permit CO_2 wells under existing EPA regulations and guidance.

Question 2

Creating an infrastructure to capture, transport, store, and monitor CO_2 will take greater federal resources including staff, technology and other elements. Do you think your agency is well-equipped to begin to undertake this enormous challenge?

Answer. There are several federal agencies that will play a role in establishing a national program to carry out carbon dioxide capture and storage (CCS) on the scale that will be needed to address climate change. While EPA itself will not undertake infrastructure projects, the Agency is responsible for implementing environmental statutes and other programs that may affect deployment of CCS. EPA has been and continues to thoroughly examine its CCS-related statutory and programmatic responsibilities in order to prioritize Agency efforts.

Along those lines, establishing a regulatory framework for geologic sequestration (GS) under the UIC program is an integral step towards creating an enabling framework for CCS. EPA is committed to continuing to provide funding and resources for regional and state regulators for the purposes of implementing the UIC program. Currently, UIC programs receive nearly $11 million annually to assist in implementing their programs.

The UIC Program has regulated over 800,000 injection wells for over 35 years. While the GS of CO_2 is a new technology that poses a unique set of risks to underground sources of drinking water and human health, EPA believes that GS can be a safe and effective tool when wells are properly sited, operated, monitored, and closed. We believe the UIC program provides an appropriate regulatory framework within which to manage the injection of CO_2 to ensure protection of underground sources of drinking water.

Question 3

Many of the introduced climate change bills have called for ''permanent'' storage, but do not go so far as to define ''permanence.'' Since the EPA may have a role in monitoring CO_2, does the agency have a clear definition of ''permanent CO_2 storage'' as it relates to geologic storage? Would there be any amount of leakage that is acceptable? What is reasonable? How can we enforce such a requirements?

Answer. Effectiveness of geologic storage is contingent on CO_2 remaining stored underground for a long period of time. A desirable timeframe for geologic storage of CO_2 is on the order of thousands of years or longer. However, effectively sequestering CO_2 for even a few hundred years could provide valuable flexibility in reducing CO_2 emissions and contribute to reducing the costs of mitigating climate change.

For the purposes of EPA's proposed rulemaking under the UIC program, we have not yet defined this term. Generally, the CO_2 will need to be stored for long periods of time (e.g., centuries or millennia) in a manner that does not endanger underground sources of drinking water.

EPA is currently deliberating on several of these issues as the Agency works to develop a regulatory proposal under the UIC program. EPA anticipates that some CCS projects will exhibit a certain amount of leakage within the subsurface, however, safeguards such as leak detection and monitoring will protect against leakage that endangers underground sources of drinking water. The Agency is considering requiring monitoring for leakage to the atmosphere as it is an indicator of potential leakage to or endangerment of underground sources of drinking water.

Question 4

Could you elaborate more about how the EPA could work with state regulations to monitor CO_2? Would there be similarities to how you monitor the UIC program?

Answer. The Safe Drinking Water Act, 42 U.S.C. 300h-1, allows States to apply to EPA for primary enforcement responsibility to administer the UIC program; those States receiving such authority are referred to as ''Primacy States.'' For UIC Class I,111, IV and V wells, states must meet EPA's federal minimum requirements for UIC programs, including minimum construction, operating, monitoring and testing, reporting, and closure requirements for well owners or operators. Where states do not seek this responsibility or fail to demonstrate that they meet EPA's federal minimum requirements, EPA is required, by statute, to implement a UIC program for such States (42 U.S.C . 300h–1(c)). We expect that states who wish to implement a CCS program would be subject to similar requirements for primacy and would need to demonstrate that they meet EPA's federal minimum requirements for CCS.

Response of Benjamin H. Grumbles to Question from Senator Menendez

Question 1

Mr. Grumbles, CCS technology does not necessarily address other environmental problems with coal fired power plants. Coal is a major source of air pollution, with coal-fired power plants spewing 59% of total U.S. sulfur dioxide pollution and 18% of total nitrogen oxides every year. Coal-fired power plants are also the largest polluter of toxic mercury pollution, the largest contributor of hazardous air toxics, and release about 50% of the nation's particle pollution. In addition, mining coal itself can pollute groundwater and devastate landscapes. Do you agree that even with an effective carbon capture and sequestration program that other environmental harms from coal need to be addressed before CCS technology can truly usher in an era of ''clean coal?''

Answer. EPA is committed to addressing environmental challenges associated with the use of coal, which is an abundant domestic energy source that is important to U.S. energy security. Under the Clean Air Act, EPA has made and will continue to make significant achievements in reducing major pollutants from coal fired power plants, including sulfur dioxide, nitrogen oxides, and mercury. Reductions in air pollution, since the 1990 Clean Air Act Amendments, have moved significantly further through policies such as the Clean Air Interstate Rule. The proposed rulemaking under the Underground Injection Control (UIC) Program is an important step towards ensuring protection of U.S. drinking water. EPA is also working under the Clean Water Act (CWA) with other Federal agencies, the States, and the coal mining industry to significantly reduce adverse impacts to the Nation's waters from surface coal mining activities. We are using our CWA regulatory tools to ensure mining impacts to streams and wetlands are avoided wherever possible, and where impacts can not be avoided, we are requiring more effective mitigation and reclamation to offset these impacts.

Responses of Ronald T. Evans to Questions from Senator Bingaman

Question 1

Many people have called for ''permanent'' storage in introduced legislation (e.g. all the introduced climate bills), but do not go so far as to define permanence. Can any one of you elaborate on a clear definition of ''permanent CO_2 storage'' as it relates to geologic storage?

Answer. The only definition or measure of permanent storage of which I am aware is that set forth by the Intergovernmental Panel on Climate Change (IPCC) which states that storage should be on the order of 1,000 years in a geologic formation, with less than 1% leakage of the volume of CO_2 that is injected.

Question 2

What is the appropriate amount of time that CO_2 should be stored in the subsurface, as means of mitigating CO_2 emissions? Will storage times of that magnitude be burdensome on CCS projects?

Answer. I am not an expert on climate change and thus not in a position to recommend the appropriate length of time CO_2 should be stored underground to mitigate emissions.

Whether or not storage times will be burdensome on CCS projects—specifically on projects undertaken in conjunction with enhanced oil recovery, Denbury's area of operations—will depend on the nature and scope of any post-injection monitoring requirements. Denbury currently monitors and verifies CO_2 volumes we inject in enhanced oil recovery projects in order to properly manage the project. When it is determined that all of the economically producible oil or gas has been recovered, the projects are generally abandoned by properly plugging the wells. However, to date, we have not sequestered any volumes of CO_2 for permanent storage and thus have no experience managing long-term monitoring requirements. Further, we are not aware of any companies, or governments for that matter, that have been in existence for 1,000 years. Thus, the 1,000 year period being suggested by the IPCC seems difficult to envision, much less manage, particularly without knowledge of what will be required to satisfy regulatory agencies overseeing CCS projects.

Question 3

Is there any amount of leakage that is acceptable? The IPCC suggests that storage should be on the order of 1000 years in a geologic formation, with less than 1% leakage of the volume of CO_2 that is injected over the life of the storage project. Is this a reasonable expectation? How can we enforce such a requirement?

Answer. I am not an expert on climate change and thus not in a position to recommend the amount of CO_2 leakage that is ''acceptable.''

It is difficult to say whether 1% leakage of the volume of CO_2 that is injected over the life of the storage project is a reasonable expectation. Based on our experience, we believe that CO_2 injected into a geological reservoir during and following the successful completion of enhanced oil recovery will remain in the ground permanently without leakage, assuming the project is abandoned properly following the recovery of the oil or gas. Subsequent events such as improper cementing, the drilling of wells through the geologic reservoir seeking additional minerals, or subsurface seismic activities, could cause the trapping mechanism to be breached, leading to some leakage. These risks can be properly managed to reduce the potential for leakage, but it is impossible to assert a particular percentage amount of leakage that is reasonable to expect from CCS projects generally.

In the short run, to enforce any leakage limitation, credits for sequestration could be withdrawn or withheld proportionate to the amount of leakage from a particular site. The best way for the government to enforce such a limit over the long run would seem to be for the government to assume control over and/or responsibility for the stored CO_2.

Question 4

Is Denbury or other EOR companies concerned that the EPA regulatory effort may be too onerous or prescriptive? Do you think they should be the lead agency in developing regulations for CCS (at the State or Federal level)?

Answer. It is difficult for Denbury to assess the EPA's regulatory effort until the EPA proposes rules for comment. Nonetheless, we are concerned about how the EPA will eventually classify CO_2 emissions, and subsequently, what requirements the EPA will propose concerning how to capture and store CO_2. CO_2 is not poisonous, explosive or flammable and has been vented for many years without incident. In addition, CO_2 is currently consumed by individuals, consumed by plants, and utilized in the refrigeration of food. Thus, classifying CO_2 as any sort of hazardous substance would be inconsistent with its safety and its current uses.

I am not certain whether EPA is the appropriate lead agency in developing regulations for CCS. As stated in my written hearing testimony, Denbury agrees with the recommendation of the Interstate Oil and Gas Compact Commission (IOGCC) that future CO_2 policies should build upon the state-based framework already in place, as this framework has provided an entirely safe development of infrastructure for EOR involving CO_2. Based on Assistant Administrator Grumbles' testimony at the hearing, the EPA plans to address the injection of CO_2 emissions for CCS in its proposed rules this coming summer. His testimony indicated EPA would be working with a variety of government agencies within the existing framework of the underground injection control regulations. We believe that in developing proposed regulations on Underground Injection Control specific to CCS, EPA should work closely with the Department of the Interior's Minerals Management Service, which has significant experience managing the nation's natural gas, oil and other mineral resources.

Question 5

Many industry professionals have indicated that for large scale CCS to take hold, government must help provide certainty by developing a legal and regulatory framework for the storage of CO_2. Besides resolving technological hurdles, what are the first steps that the government can take specific to legal, regulatory or social elements that will allow more CCS projects to go forward?

Answer. First, Congress should amend tax code Section 7704 to clarify that subsection (d)(1)(E) covers man-made, as opposed to just naturally occurring, CO_2. With the current legal uncertainty in Sec. 7704, much of the existing CO_2

pipeline capacity cannot be used—and new capacity may not get built—to transport anthropogenic CO_2. This action would not be providing a new incentive for CCS; it would simply remove existing legal ambiguity.

Second, as discussed in my written testimony, federal eminent domain authority may ultimately be required to develop a nationwide CO_2 pipeline infrastructure.

Third, we believe that the current regulatory structure is sufficient with respect to (a) CO_2 pipeline safety, overseen by the Office of Pipeline Safety in the U.S. Department of Transportation's Pipeline and Hazardous Materials Safety Administration, (b) CO_2 pipeline access, which is not currently federally regulated and need not be, due to the small, regional, non-integrated nature of existing and planned CO_2 pipelines, and (c) protection of the fresh water and ground water reservoirs from the injection of CO_2, as provided in the Safe Drinking Water Act and by the EPA's Underground Injection Control Program.

Finally, we concur with the IOGCC's recommendation that future CO_2 regulation should build upon the primarily state-based regulatory framework already in place, due to the states' decades of experience with CO_2 EOR, natural gas storage, and acid gas injection.

Responses of Ronald T. Evans to Questions from Senator Dorgan

Question 1

I also think it's important that the general public have an understanding of how vastly important an issue this will be to our energy future. In your judgment, what will it take for the general population to better understand and support the approaches associated with carbon capture and storage?

Answer. For the general population to better understand and support CCS, it is important to educate and disseminate information about various approaches to CCS, such as EOR, through various forums and Congressional hearings similar to the one held by this committee. The general public and legislators need to be educated about the various costs, merits and limitations of various energy sources, including the costs of CCS, available to the country along with reasonable forecasts of the energy needs of this country.

Capturing and storing CO_2 from various sources is technically feasible today, can provide additional sources of energy through enhanced oil recovery, and assist in the reduction of greenhouse gases. The transportation and sequestration of CO_2 related to enhanced oil recovery has been safely demonstrated for over 30 years with no loss of life due to CO_2 leakage. Particularly important is the necessity to distinguish between the relative safety of CO_2 in comparison to the other materials transported via pipeline such as oil, natural gas, petroleum products and hydrogen. CO_2 is not poisonous, explosive or flammable and thus poses far less risk when transported.

Question 2

I have seen reports that indicate that over 200 billion barrels of oil may remain as residual oil in geologically complicated, partially-produced or in mature oil fields in the U. S. What can you tell me about the potential for oil recovery using CO_2 EOR (enhanced oil recovery) that is practiced today versus the potential for so called ''next generation'' EOR technologies?

Answer. The various technologies associated with EOR (e.g. water flooding, steam flooding, CO_2 flooding, polymer flooding, advanced well designs, etc.) have been used and developed over the past thirty years. The CO2flooding used by Denbury today is state of the art and the most efficient at recovering oil in the proper applications. Nonetheless, we continue to refine and improve EOR processes in an attempt to yield additional barrels of oil. Future technological advances will likely allow recovery of incremental volumes of oil, rather than result in a quantum leap or step change in amounts recoverable.

In general, less than 50% of the oil discovered in this country was or will everbe produced, and thus a significant volume of oil is still in the ground and unable to be recovered without some sort of additional investment. The U.S. Department of Energy commissioned a study of the potential amount of oil that could be recovered from CO_2 enhanced oil recovery. The results of this study, completed by Advanced Resources International in 2006, indicated that approximately 390 billion barrels of the 580 billion barrels of oil originally discovered will remain in theground after primary and secondary methods are applied. Using current technologies it is technically possible to recover approximately 89 billion barrels via CO_2 enhanced oil recovery. These technically recoverable barrels are based on several assumptions within the report concerning CO_2 prices, oil prices and other variables. The report can be found at: http://www.fossil.energy.gov/programs/oilgas/eor/Ten_Basin-Oriented_CO2-EOR_Assessments.html

Question 3

Many industry professionals have indicated that for large scale CCS to take hold, federal and state governments must help provide certainty by developing a legal and regulatory framework for the storage of CO_2. Besides resolving technological hurdles, what are the first steps that the government can take specific to legal, regulatory or societal elements that will allow more CCS projects to go forward?

Answer. Please see my response to Senator Bingaman's question #5. In addition, the Coal Fuels and Industrial Gasification Demonstration and Development Act, S. 2149, which you introduced on October 4, 2007, includes the necessary clarification of Section 7704(d)(1)(E) to cover man-made CO_2 and Denbury commends your leadership on this issue.

Question 4

I believe that in the near term we need to find ways to create certainty for investors to deploy CCS projects. I introduced a bill called the Clean Energy Production Incentives Act (S. 1508) earlier this year that among other things includes a 10-year production tax credit for the storage of CO_2. Developers receive a higher credit for permanent storage and slightly lower credit for using enhanced oil recovery techniques. The bill also included accelerated depreciation and tax credit bonds for CO_2 capture and storage property. Can you talk more about the scale of incentives and how these incentives can accelerate development of large scale carbon capture and storage? Are the incentives adequate to incentivize near-term and long-term storage options?

Answer. I agree that we need to find ways to create certainty for investors to deploy CCS. The production tax credits and accelerated depreciation of pipelines proposed in the Clean Energy Production Incentives Act introduced by you are meaningful incentives to encourage the development of large scale CCS. The proposed level of incentives should be sufficient to encourage the capture of the lowest cost emissions currently and a significant number of proposed gasification projects. Our analysis of the costs to capture, transport and store the lowest cost emissions indicates total costs of approximately $20/ton. Thus a $20/ton tax credit should be meaningful to projects that will be required to sequester their emissions in deep saline reservoirs. The corresponding lower amount, $10/ton for CO_2 sequestered in EOR projects, is sufficient to encourage the development of additional enhanced oil recovery projects using volumes of captured anthropogenic CO_2. These tax credits would encourage the development of CCS projects in the near term.

Technological solutions will have to be developed in coming years to decrease the costs of CCS for many additional existing sources of CO_2 emissions.

The cost of CCS for many of these sources is up to $70/ton and potentially higher. The current proposed production tax credits are not sufficient to encourage or incentivize the capture and storage of these emissions. Necessary technology will only be developed, and the costs of CCS driven down for these sources, by getting the lowest cost CCS projects up and operating. As the lowest cost emissions get captured and research and development progresses, we are convinced that technological and cost break-throughs will occur. However, delay in incentivizing the lowest cost CCS projects will delay the overall timeframe of the technology development necessary to ultimately address the higher cost emissions.

Response of Ronald T. Evans to Question from Senator Menendez

Question 1

You claim that government incentives are necessary to promote the use of man-made carbon dioxide in enhanced oil recovery [EOR] projects. I find this claim troubling on a number of levels. EOR was economically viable even back when we had cheap oil. Why isn't $90 a barrel oil all the incentive needed? In fact, there are several existing projects which do in fact capture man-made carbon dioxide for use in EOR. The process is economically viable right now, without any government incentives. More importantly, using CO_2 to recover more oil won't stop global warming because we may be putting some CO_2 in the ground, but the extra oil we extractwill lead to more emissions of CO_2 when that oil is used. Why should EOR operations get credit for reducing greenhouse gas emissions, when the process is probably a wash at best in terms of the reduction of net CO_2 reductions?

Answer. In my written testimony, I stated that ''without some means of reducing the cost of capturing anthropogenic CO_2 significantly, infrastructure development will likely remain stagnant.'' As explained therein, the cost of capturing and compressing anthropogenic CO_2 is significant. EOR can indeed absorb a portion of the capture and compression costs, but the percentage of the capture and compression costs that EOR can cover is dependent upon the level of oil prices. The total cost of CCS from different industrial sources varies widely from $20/ton to over $70/ton. EOR can cover the costs of transportation and sequestration, which have seen significant increases over the last few years, and cover a portion of the capture and compression cost. Unfortunately, even with oil at $90 per barrel, EOR cannot cover 100% of the CCS costs for a significant number of current emitters.

Naturally occurring CO_2 is generally priced as a percentage of oil prices. Thus, when oil prices were low, CO_2 prices were reduced in order to keep projects economic, but at very low oil prices almost all of the projects were uneconomic. As prices have increased, the cost of naturally occurring CO_2 has also increased. In addition to increasing CO_2 costs, the capital costs associated with CO_2 facilities and well work have increased substantially and in some cases the costs have more than doubled.

The only significant use of man-made CO_2 in EOR at present is EnCana Corporation's use of CO_2 from the North Dakota Gasification Project, which was made possible with significant support from the Canadian government. The other anthropogenic sources of CO_2 currently captured and used in EOR in the United States, primarily in the Permian Basin of West Texas, are produced along with natural gas from underground reservoirs. The reason this CO_2 is available at a relatively low cost is because the CO_2 must be separated from the natural gas in order to sell the natural gas. This separation results in a relatively pure stream of CO_2 that then only has to be compressed and transported short distances to the oil field.

The reason EOR should qualify for whatever credit may eventually exist for reducing green house gas emissions is straightforward: First, America will continue to rely on fossil fuels to meet a substantial portion of its energy needs for the foreseeable future and a molecule of CO_2 stored, whether in EOR or non-EOR sequestration, is a molecule that will not get vented to the atmosphere. Second, use of EOR to produce oil domestically avoids the significant emissions and potential spills associated with shipping oil from overseas. Third, according to a recent study of CO_2 storage in connection with EOR commissioned by the U.S. Department of Energy's National Energy Technology Laboratory, advances in technology could enable storage of 1.6 times as much CO_2 in oil reservoirs as the CO_2 content in the recovered oil. (See: http://www.netl.doe.gov/energy-analyses/ pubs/ Storing%20 CO2% 20w% 20 EOR_FINAL.pdf) At present, EOR is the only viable process that can accept and use productively large volumes of anthropogenic CO_2 that would otherwise be emitted.

In addition, EOR benefits U.S. energy security by producing additional volumes of oil in the United States and displacing the necessity for imports from foreign countries. The barrel of oil produced by U.S. workers in the U.S. is worth significantly more in terms of domestic investment, jobs and energy security than the value of an imported barrel of oil.

Keeping up with America's evergrowing demands for energy while reducing emissions of greenhouse gases will take an enormous effort across all fields of energy development, all manner of industry, and all levels of society. If the twin goals of energy independence and a cleaner environment are to have any hope of being achieved without significant reductions in our standard of living, the country cannot afford to ignore particular resources or technologies; the U.S. must utilize every means available to it. EOR is not the answer to all of America's energy and emissions challenges, but it is one of the only readily available alternatives that is working now and that can be broadly utilized in the near term if policymakers decide action is imperative.

Responses of Krista L. Edwards to Questions from Senator Bingaman

Question 1

Many people have called for ''permanent'' storage in introduced legislation (e.g. all the introduced climate bills), but do not go so far as to define permanence. Can any one of you elaborate on a clear definition of ''permanent CO_2storage'' as it relates to geologic storage?

Answer. I would respectfully defer to the U.S. Department of Energy and otheragencies represented at the hearing. The U.S. Department of Transportation's Pipe-line and Hazardous Materials Safety Administration (PHMSA), through its two safety programs, prescribes and enforces standards for the design, testing, and maintenance of tanks, cylinders, and other containers used in the transportation and incidental storage of CO_2. But we do not regulate any aspect of the long-term, permanent, or geologic storage of CO_2 and, accordingly, have had no occasion to consider the requirements for permanent geologic CO_2 storage.

Question 2

What is the appropriate amount of time that CO_2 should be stored in the subsurface, as means of mitigating CO_2 emissions? Will storage times of that magnitude be burdensome on CCS projects?

Answer. I would respectfully defer to other agencies represented at the hearing. As I mentioned in response to an earlier question, PHMSA's oversight extends tohazardous materials transportation, including incidental storage, and we have had no occasion to consider the requirements for long-term geologic storage of hazardous materials.

Question 3

Is there any amount of leakage that is acceptable? The IPCC suggests that storage should be on the order of 1000 years in a geologic formation, with less than 1% leakage of the volume of CO_2 that is injected over the life of the storageproject. Is this a reasonable expectation? How can we enforce such a requirement?

Answer. I would respectfully defer to the other agencies represented at the hearing. As mentioned in response to an earlier question, PHMSA has had no occasion to consider standards for permanent or long-term geologic storage.

Question 4

In your testimony you stated that CO_2 pipelines have a strong safety record. Does your agency do any sort of public outreach to convey this message tothe general public?

Answer. Yes. PHMSA regularly posts safety-related data, including incident statistics, on our public website (http://primis.phmsa.dot.gov/comm /reports/safety/ PSI.html). In addition, PHMSA has several program initiatives focused on safety-related public education and outreach. The agency's Community Assistance and Technical Services (CATS) program is devoted to enhancing public understandingabout pipeline operations and risk controls through direct interaction with local officials and concerned citizens and the development and dissemination of training programs. CATS representatives working out of the agency's five regional offices participate directly in community meetings and public hearings related to existing or proposed pipelines. As experienced pipeline engineers, our CATS representatives are uniquely qualified to answer public questions and concerns about pipeline operations and oversight.

As part of a comprehensive approach to pipeline safety, PHMSA also requires pipeline operators to develop and implement system-specific public awareness programs targeting nearby communities. These programs are designed to enhance public understanding about pipeline risks and safety performance and to educate communities about the prevention, detection, and reporting of pipeline events. Together with our state partners, PHMSA reviews operators' public awareness programs and enforces requirements under which operators must periodically evaluate their programs' effectiveness in reaching targeted populations and satisfying information needs.

Building on these efforts, PHMSA's Pipelines and Informed Planning Alliance (PIPA) is bringing together a broad group of stakeholders (including industry, safety advocates, and state and local officials) to promote the development of risk-informed standards to guide land use and community planning. We launched PIPA in January of 2008 and have arranged working group meetings throughout the year, targeting a final report by January 2009.

Question 5

Do you envision PHMSA taking on an expanded regulatory role should the existing CO_2 pipeline network be expanded beyond its current geographic distribution? Are there any areas where CO_2 pipelines may be more challenging to site and regulate?

Answer. PHMSA's existing pipeline safety program has provided effective oversight of CO_2 pipelines since 1991 and will accommodate new CO_2 pipelines, however and wherever they are located. From the perspective of public safety and environmental protection, the CO_2 pipeline network can be expanded beyond its current geographic distribution using existing technologies and under existing safety standards and oversight arrangements.

Under PHMSA's current program, a limited number of states are certified to oversee hazardous liquid pipelines, including CO_2 pipelines. To the extent that new CO_2 pipelines are planned in other states, we will be prepared to work with state and local officials to address information requirements and, as appropriate, help states expand their program certifications to encompass CO_2 pipelines.

I would defer to other witnesses concerning the economic and regulatory challenges associated with siting of CO_2 pipelines.

Responses of Krista L. Edwards to Questions from Senator Dorgan

Question 1

It seems to me that we need to much more quickly begin establishing and defining the ''rules of the road'' when it comes to carbon management. As we begin to unlock the opportunities for capturing, moving and storing larger amounts of CO_2, it is fair to say that the federal government will likely play a greater role. It will be better if we begin to better define appropriate roles for local, state and federal governments.What are the most critical near-term issues that your agencycan address so that developers can begin demonstrating CCS projects?

Answer. PHMSA and our state partners currently oversee nearly 4,000 miles of CO_2 pipelines under established and effective standards for public safety and environmental protection. The same oversight arrangements and standards will govern new CO_2 pipelines, however and wherever they are configured and located. Accordingly, PHMSA foresees no significant challenges meeting its statutory responsibilities with respect to CCS projects involving pipeline transportation. We will be prepared to work with communities, prospective operators, and state pipeline safety programs to plan for the construction, operation, and oversight of new CO_2 pipelines.

Under our current program, a limited number of states are certified to oversee hazardous liquid pipelines, including CO_2 pipelines. To the extent that new CO_2 pipelines are planned in other states, we will be prepared to work with state and local officials to address information requirements and, as appropriate, help states expand their program certifications to encompass CO_2 pipelines.

Question 2

Creating an infrastructure to capture, transport, store and monitor CO_2 will take greater federal resources including staff, technology and other elements. Do you think your agency is well-equipped to begin undertaking this enormous challenge?

Answer. Yes. Having successfully overseen the operation of CO_2 pipelines since 1991, DOT is well-positioned to carry out its statutory responsibilities with respect to future CO_2 transportation. PHMSA's existing pipeline safety programs and standards are established and effective, as reflected in accident trends generally and the strong safety record of the roughly 4,000 miles of CO_2 pipelines currently in operation. Since we introduced our Integrity Management program in 2000, the annual number of serious incidents involving hazardous liquid pipelines has reached historic lows, even as the size of the pipeline network has grown. Within these data, the safety record of CO_2 pipelines is especially strong, with no incidents involving death or serious injury since the inception of DOT oversight in 1991.

Together with our state partners, PHMSA currently oversees more than two million miles of natural gas and hazardous liquid pipelines, including pipelines carrying crude oil, refined products, highly volatile liquids, anhydrous ammonia and hydrogen, in addition to carbon dioxide. Because it is neither combustible, reactive, nor toxic, CO_2 is the least hazardous of the materials regulated under our pipeline safety program.

The Nation's pipeline network is expanding, with numerous privately-financed pipeline construction projects currently underway and in the planning and permitting stages. Although I would defer to other members of the panel for more precise data and projections, we expect this growth to continue, without regard to any pipe-line expansion associated with CCS projects.

In accordance with the Pipeline Inspection, Protection, Enforcement, and Safety Act of 2006 and our FY 2008 appropriations, PHMSA is in the process of increasing the numbers of federal inspection and enforcement personnel and increasing financial support for state pipeline safety programs. The President's FY 2009 budget proposes further increases in staffing and state grants, addressing a variety of new oversight demands, including the construction of new and expanded energy pipe-lines and the use of existing pipelines for the transportation of new fuel products and blends.

PHMSA's existing program can accommodate the expansion of CO_2 pipelines, without significant new challenges or resource requirements. The current CO_2 pipeline network accounts for roughly two percent of the hazardous liquid pipeline mileage under our jurisdiction (and less than 0.2 percent of the total pipeline mileage we oversee). Accordingly, even a significant expansion of the CO_2 pipeline network will only marginally increase the mileage under our jurisdiction and is unlikely to necessitate any changes in our standards and oversight arrangements. PHMSA is prepared to work with communities and prospective operators to plan for new CO_2 pipelines, and we will be prepared to oversee their construction and operation, just as PHMSA is doing today in connection with the expanding network of natural gas,crude oil, and refined products pipelines.

As I mentioned in response to an earlier question, PHMSA anticipates that state pipeline safety programs may play a larger role in oversight of new CO_2 pipelines, depending on the location and configuration of the new lines. We will be prepared to provide financial and technical support to states interested in participating in oversight of CO_2 pipelines.

Responses of Lawrence E. Bengal to Questions from Senator Bingaman

Question 1

Many people have called for ''permanent'' storage in introduced legislation (e.g. all the introduced climate bills), but do not go so far as to define permanence. Can any one of you elaborate on a clear definition of ''permanent CO_2 storage'' as it relates to geologic storage?

Answer. Using the term ''permanent'' when referring to storage of carbon dioxide (CO_2) unnecessarily identifies CO_2 with the disposal of manmade ''wastes'', which essentially remain the same in the subsurface, unlike CO_2. CO_2 is a ''natural'' gas that, in addition to being part of the air we breathe, can be found naturally in the subsurface throughout the world. Naturally stored CO_2 is part of the earth's geologic processes. Injected CO_2 will likewise become part of the natural earth processes, mineralizing, going into solution and eventually reaching a stability as do natural CO_2, natural gas and crude oil reservoirs, in which CO_2, natural gas and oil are ''stored'' naturally for millennia.

It is important also to realize that it is primarily during the active injection phase that there is a risk of leakage into the atmosphere. These risks can be significantly reduced if not eliminated by injecting the CO_2 only into select geologically sound storage reservoirs and making certain through subsurface monitoring that the CO_2 during the active injection phase is being contained within the storage interval. Once the injection ceases and the injection facility is closed, the risks of migration from the reservoir decrease exponentially as the CO_2 incorporates with the reservoir environment. Should a leak be detected during the injection phase, remedial actions can be initiated to address the leak until the situation is mitigated. Although monitoring will continue during the post closure phase, it unlikely to reveal leakage that hadn't been detected during and immediately after the active injection phase.

In summary, at the time a site is closed, the regulator would be expected to have a good understanding of how the CO_2 is behaving in the storage interval and a good deal of confidence that the injected CO_2 is likely to remain in that interval over the very long term. As I have discussed, however, the term ''permanence'' seems to be more applicable to the situation where a non-natural man-made substance is being emplaced into the earth. In this case, in a very real sense, CO_2 becomes part of the geologic environment into which it is emplaced. It is thus not so much ''permanently'' contained, as it is incorporated, albeit permanently, into the geologic system as are natural accumulations of oil and natural gas.

Question 2

What is the appropriate amount of time that CO_2 should be stored in the subsurface, as means of mitigating CO_2 emissions? Will storage times of that magnitude be burdensome on CCS projects?

Answer. The answer to the prior question basically answers this question as well. If all goes as anticipated in the storage of CO_2 in an appropriate geologic reservoir, where during the injection, closure and post-closure phases no leakage from the storage interval has been detected by monitoring, it can generally be assumed that the CO_2 has become integrated into the geology of the interval into which it has been emplaced. As previously discussed, there is no reason to anticipate post closure leakage (see also the answer to question #3), although monitoring will continue in-definitely. The IOGCC-proposed state administered trust fund, to which operation of the site would be transferred post closure, would be designed to be more than adequate to indefinitely monitor the site and take, should such be deemed necessary, any and all remedial actions. The specification of a ''storage time'' would not be a burden, but an expected part of the post closure regime. Of more importance would be the definition of ''leakage'' and the regulatory burden a strict ''unmeasurable'' standard would impose on a CO_2 project (see next question).

Question 3

Is there any amount of leakage that is acceptable? The IPCC suggests that storage should be on the order of 1000 years in a geologic formation, with less than 1% leakage of the volume of CO_2 that is injected over the life of the storage project. Is this a reasonable expectation? How can we enforce such a requirement?

Answer. In answering this question it is important to be clear what we are trying to accomplish in regulation of the storage of CO_2. We begin with the assumption that first and foremost the reason we are geologically storing CO_2 in the first place is in lieu of ''storing'' (releasing) this greenhouse gas in the atmosphere where it willcontribute to global climate change.

There are three regulatory standards which will determine an acceptable amount of ''leakage''; 1) carbon credits—the leakage threshold in this respect will likely be established by the governmental and or commercial entity which is ultimately established to award carbon credits for the underground storage of CO_2; 2) public health and safety—while there is a threshold of even minimal release into the atmospherewhich could endanger public health and safety, catastrophic release from underground storage is almost inconceivable, and; 3) drinking water protection—this is the purpose of the regulations being developed by the U.S. Environmental Protection Agency (EPA), likely to be administered by states under delegated authority from EPA under its Underground Injection Control program. Presumably these regulations will establish this threshold, which is likely to be ''no contamination''.

Having said this, it is very clear that the most important factor in assuring that stored CO_2 remains in the subsurface is to make certain that the geologic sites selected for the storage of CO_2 are optimal for that purpose and mirror the ''storage'' capabilities of naturally occurring oil and natural gas reservoirs. I therefore begin with the assumption and expectation that CO_2 will be stored in formations for which the geology indicates a very strong likelihood of long term containment which meet the thresholds of all three regulatory interests identified above. I further assume and fully expect that the regulatory framework which permits any such storage facility will contain layers of safeguards designed to prevent leakage. However, the purpose of subsurface monitoring is, first and foremost, to detect leakage in the sub-surface (above the storage interval) well before the CO_2 reaches the surface and hence the atmosphere. However, even in the very unlikely event that monitoring reveals leakage and that leak is mitigated (mitigation options range from complete cessation of injection—including ultimately depressurizing of the storage interval until leakage is no longer occurring—to capture and re-injection of the leaking CO_2 all while still contained in the subsurface) I am assuming and expecting that any escaping CO_2 will be unlikely to reach the surface and hence the atmosphere.

Question 4

In your testimony you outline the operational bond requirements for the post-closure phase of the CCS project. Following site closure, the operational bond is released the regulatory liability is turned over to a trust fund administered by the state. Did the IOGCC consider a period of time whereby private insurance companies could manage these sites (post-closure)?

Answer. No. Based on the states' long experience with financial assurances in the plugging of oil and gas wells and the administration of state administered abandoned oil and gas well programs, the IOGCC Task Force believed that the state-administered trust fund would offer the greatest flexibility post closure to monitor and ''caretake'' the facility in perpetuity. Private insurance would lack the flexibility and responsiveness to be able to immediately respond to potential contingencies. It would also require that a regulator adopt inflexible rules setting the parameters under which insurers would operate.

Responses of Lawrence E. Bengal to Questions from Senator Dorgan

Question 1

I also think it's important that the general public have an understanding of how vastly important an issue this will be to our energy future. In your judgment, what will it take for the general population to better understand and support the approaches associated with carbon capture and storage?

Answer. I couldn't agree more. I indicated in my testimony the importance of public support for this technology, but before the public can support this technology, they must understand and feel ''comfortable'' with the process. The U.S. Department of Energy has understood this and has required each of the Regional Carbon Sequestration Partnerships to have a strong public outreach component. It has also funded the Interstate Oil and Gas Compact Commission to widely distribute the findings of the Task Force's work regarding geologic storage of CO_2. The most recent publication of the IOGCC entitled Road to a Greener Energy Future— CO_2 Storage: A Legal and Regulatory Guide for States is being circulated broadly and can be downloaded at the IOGCC website at (http://www.iogcc.state.ok.us/ PDFS/Road-to-a-Greener-Energy-Future.pdf).

We also firmly believe that a key component of building public support is making absolutely certain that states in their process of creating the laws and regulations to govern this technology do so in a completely open and transparent manner. It is essential that all stakeholders be included in the process if the public is to have confidence in the technology. The public must understand that site selection will be a very important part of the regulatory process. Not every site will be suitable for storage. Only the sites most geologically suitable will be considered.

Also important will be how this issue is presented to the public. As I indicated in my testimony, the public needs to be informed about the long history of CO_2 transportation, handling and use in a great variety of applications. They need to understand that CO_2 is a substance that is part of the air we breathe and that storing it underground is not something entirely new and mysterious, but the technological outgrowths of things with which states already have regulatory experience, like oil and natural gas development, natural gas storage, and CO_2 injection for enhanced oil recovery. I believe it will be important that the public understand that the production of CO_2 is a consequence of the public's demand for and use of fossil energy and that it is arguably in the public's interest to actively participate along with industry in efforts to reduce CO_2 emissions through geologic storage.

Given the regulatory complexities of CO_2 storage including environmental protection, ownership and management of the pore space, maximization of storage capacity and long term liability, geologically stored CO_2 should be treated under resource management frameworks as opposed to waste disposal frameworks.

Regulating the storage of CO_2 under a waste management framework sidesteps the public's role in both the creation of CO_2 and its interest in the mitigation of its release into the atmosphere and places the burden solely on industry to rid itself of "waste" from which the public must be "protected". Such an approach lacking citizen buyin with respect to responsibility for the problem as well as the solution could well doom geological storage to failure and diminish significantly the potential of geologic carbon storage to meaningfully mitigate the impact of CO_2 emissions on the global climate.

A resource management framework, as proposed by the Task Force, allows for the integration of these issues into a unified regulatory framework and proposes a "public sector-private sector partnership" to address the long-term liability, given that the release of CO_2 into the atmosphere is at least partially a societal problem and that the mitigation of that release is likewise at least partially a societal responsibility and clearly a societal benefit.

Question 2

Many industry professionals have indicated that for large scale CCS to take hold, federal and state governments must help provide certainty by developing a legal and regulatory framework for the storage of CO_2. Besides resolving technological hurdles, what are the first steps that the government can take specific to legal, regulatory or societal elements that will allow more CCS projects to go forward?

Answer. Echoing a point made above and in my testimony, the technology which will be used to store CO_2 underground is not something new, but a technological outgrowth of things with which states already have regulatory experience, like oil and natural gas development, natural gas storage, and CO_2 injection for enhanced oil recovery. What could kill this technology is regulatory uncertainty in such areas as long term liability. If the federal government imposes a Superfund model on storage, the technology is probably dead on arrival as a viable means of mitigating the impacts of climate change on the environment. If on the other hand investors and companies interested in undertaking the storage of CO_2 are reasonably assured that a long term regulatory system along the lines of that proposed by the IOGCC Task Force is adopted, then such an atmosphere will likely be much more conducive to rapid development of the technology.

Other Issues of Importance

The IOGCC team clearly understands that there are large numbers of underground sites with effectively no risk of leakage and other sites where CO_2 injection should not be permitted. States, working with Federal officials, need to develop a set of criteria by which the optimal sites are chosen first. Such things as properties of primary seals, the presence of secondary and tertiary seals clearly need to be considered in such a ranking.

Aggregation of sufficient mineral and/or storage rights will be a large obstacle for this activity in a large number of places within otherwise suitable areas for sequestration. Both Federal and State rules may need clarification and action to facilitate such aggregation activities.

Many areas with CO_2 emissions will be effectively unsuitable for sequestration. Pipelines moving CO_2 from such areas to suitable ones will therefore be necessary. State and Federal legislation could simplify this task. Adoption of an openaccess status to those pipelines may be appropriate as well. Additionally, it appears that some sort of public assistance might prove a useful incentive to a private organization seeking to build a private pipeline. Incentives could come in the form of special tax treatment, access to eminent domain provisions, bonding assistance (similar to the Wyoming or North Dakota Pipeline Authorities), or some private-public arrangement (with the public part designated as open access).

Avoidance of unnecessary and ominous ‘‘permanence’’ requirements is also essential.

The post closure transfer of responsibility to a governmental entity is considered a necessity. Qualifying a site as low risk and allowing a rapid transfer in a short post-closure period would be both appropriate and stimulating for early action.

Question 3

I believe that in the near term we need to find ways to create certainty for investors to deploy CCS projects. I introduced a bill called the Clean Energy Production Incentives Act (S. 1508) earlier this year that among other things includes a 10–year production tax credit for the storage of CO_2. Developers receive a higher credit for permanent storage and slightly lower credit for using enhanced oil recovery techniques. The bill also included accelerated depreciation and tax credit bonds for CO_2 capture and storage property. Can you talk more about the scale of incentives and how these incentives can accelerate development of large scale carbon capture and storage? Are the incentives adequate to incentivise near-term and long-term storage options?

Answer. Until government imposes a limitation on the amount of carbon that can be released into the atmosphere, presumably through adoption of a carbon tax or a cap and trade system, no additional incentives are likely to prove particularly efficacious. After that happens, all of the incentives noted in the question would be most beneficial indeed. I would caveat that I would question the rationale for disadvantaging enhanced oil recovery (EOR). Perhaps the solution is to frame any production tax credit so as to advantage the CO_2 capturing entity or the pipeline entity or both rather than the injector. That effectively makes the CO_2 less expensive thus encouraging the sequestration with expanded EOR projects.

Additionally, the biggest issue today is the cost of separating and compressing the CO_2 at the source so whatever can be done on that front will be money well invested.

Responses of Joseph T. Kelliher to Questions from Senator Bingaman

Question 1

Many people have called for ‘‘permanent’’ storage in introduced legislation (e.g. all the introduced climate bills), but do not go so far as to define permanence. Can any one of you elaborate on a clear definition of ‘‘permanent CO_2 storage’’ as it relates to geologic storage?

Answer. I note, as an initial matter, that the Commission has extensive expertise in the storage of natural gas in underground reservoirs. However, the storage of natural gas differs significantly from the ‘‘permanent’’ sequestration of CO_2. Natural gas storage is a dynamic process whereby gas is injected, stored and withdrawn with some regularity while the point of CO_2 sequestration is to ‘‘permanently’’ inject and sequester the CO_2 in the reservoir.

Permanent sequestration would appear to contemplate CO_2 remaining in place for the long term. ‘‘Permanent’’ sequestration may pose problems similar to those experienced by natural gas storage reservoirs which are not cycled (the process of injecting and withdrawing gas) on a regular basis. In the case of a natural gas storage field that does not experience sufficient cycling, and does not have well defined geologic boundaries, gas could migrate through the storage formation into other formations or, in the case of aquifer storage facilities, the gas eventually could dissolve into the water and the water eventually could fill the reservoir. If a natural gas storage reservoir is cycled, and the physical infrastructure (piping, casing, surface facilities) is maintained, there is no reason to believe that the storage field could not operate indefinitely.

Question 2

What is the appropriate amount of time that CO_2 should be stored in the subsurface, as means of mitigating CO_2 emissions? Will storage times of that magnitude be burdensome on CCS projects?

Answer. The Commission has no information regarding the amount of time that CO_2 needs to be sequestered to ameliorate the effects of climate change. However, a study by the Massachusetts Institute of Technology (MIT) titled, ''The Future of Coal'' states, at p. 44, ''...it is very likely that the fraction of stored CO_2 will be greater than 99% over 100 years, and likely that the fraction of stored CO_2 will exceed 99% for 1000 years.'' (footnote omitted).

Question 3

Is there any amount of leakage that is acceptable? The IPCC suggests that storage should be on the order of 1000 years in a geologic formation, with less than 1% leakage of the volume of CO_2 that is injected over the life of the storage project. Is this a reasonable expectation? How can we enforce such a requirement?

Answer. References to leaks in natural storage fields generally apply to leaks inthe equipment at the storage field including well casings, piping, compressors, and other miscellaneous surface facilities, not the storage formation, per se. This lost and unaccounted-for gas varies by facility. Gas can also migrate through rock formations and this may be what the referenced IPCC report refers to. Natural gas storage operators drill and maintain observation wells to monitor the migration of natural gas out of the designated boundaries of their storage field. Migration out of designated storage boundaries could occur if the operator does not have a complete understanding of the geology of the storage location. This is not to suggest that the operator is at fault—despite advances in technology, it is still impossible to know exactly what lies beneath the earth's surface. However, migration can also occur as a result of improper operations or as a result of third party operations outside the boundaries of the storage field.

The MIT study referenced in the previous response addresses CO_2 storage leakage as follows:

- Importantly, CO_2 leakage risk is not uniform and it is believed that most CO_2 storage sites will work as planned. However, a small percentage of sites might have significant leakage rates, which may require substantial mitigation efforts or even abandonment. It is important to note that the occurrence of such sites does not negate the value of the effective sites. However, a premium must be paid in the form of due diligence in assessment to quantify and circumscribe these risks well. [p. 50] (footnote omitted).
- Even though most potential leaks will have no impact on health, safety, or the local environment, any leak will negate some of the benefits of sequestration. However, absolute containment is not necessary for effective mitigation. If the rate and volume of leakage are sufficiently low, the site will still meet its primary goal of sequestering CO_2. The leak would need to be counted as an emissions source as discussed further under liability. Small leakage risks should not present a barrier to deployment or reason to postpone an accelerated field-based RD&D program. This is particularly true of early projects, which will also provide substantial benefits of learning by doing and will provide insight into management and remediation of minor leaks. [p. 51] (footnotes omitted).

Question 4

In your testimony you mention three models for governing interstate pipelines. Are you prepared, as an agency, to implement any of these three models for pipeline management?

Answer. In my testimony, I referenced the oil pipeline model, where the states have siting authority, the Commission sets rates, and the Department of Transportation handles safety matters; the natural gas pipeline model, in which the Commission authorizes siting and sets rates, while the Department of Transportation handles post-construction safety; and the current carbon dioxide pipeline model, under which pipelines are sited under state law, and rates are set by the pipelines, subject to review by the Department of Transportation's Surface Transportation Board in the event a complaint is filed. Should Congress so require, the Commission could implement either of the first two models, while the third model would have no role for the Commission. However, as I said in my testimony, I would not recommend that Congress preempt the states or alter the Department Transportation's rate-making and safety roles by providing for exclusive and preemptive FERC siting of carbon dioxide pipelines, because the precondition that led Congress to such a course for siting natural gas pipelines—the failure of state siting—does not exist.

Question 5

While you state in your testimony that you feel the siting of CO_2 pipelines should stay in the jurisdiction of the state agencies—do you think that they are the appropriate entity to regulate transportation rates? This seems like it is well within the jurisdiction of FERC to regulate fair and equitable transportation costs.

Answer. As I mentioned in my written testimony, rates for carbon dioxide pipe-lines are currently set by the pipelines themselves, subject to the authority of the Department of Transportation's Surface Transportation Board to hold proceedings regarding the reasonableness of the rates, if a complaint is filed. While the Commission does possess ratemaking expertise, I have seen nothing to indicate that the current model is not functioning well. I do not recommend granting ratemaking authority over carbon dioxide pipelines to the individual states, because of the prospect of a patchwork of inconsistent rate regulation for an interstate pipeline network.

Responses of Joseph T. Kelliher to Questions from Senator Dorgan

Question 6

It seems to me that we need to much more quickly begin establishing and defining the ''rules of the road'' when it comes to carbon management. As we begin to unlock the opportunities for capturing, moving and storing larger amounts of CO_2, it is fair to say that the federal government will likely play a greater role.It will be better if we begin to better define appropriate roles for local, state and federal government. What are the most critical near-term issues that your agency can address so that developers can begin demonstrating CCS projects?

Answer. Given that the Commission has no jurisdiction over carbon dioxide pipe-lines or sequestration facilities, I do not believe that there are issues that the

Commission can address directly to speed the development of these projects. However, should Congress pass legislation such as S.2144, my staff and I would be pleased to work with other agencies to study the feasibility of carbon dioxide capture, transportation, and sequestration projects. As I said in my testimony, FERC can play a helpful role examining regulatory barriers and regulatory options relating to the construction and operation of carbon dioxide pipelines.

Question 7

Creating an infrastructure to capture, transport, store, and monitor CO_2 will take greater federal resources including staff, technology and other elements. Do you think your agency is well-equipped to begin undertake this enormous challenge?

Answer. Should Congress establish a role for the Commission in this area, I am confident that the Commission could carry it out. Depending on the nature of the responsibilities given to the Commission, and the extent of the CO_2 capture, pipeline and storage network, the Commission might need additional personnel resources were it assigned to regulate these activities.

Question 8

Your agency has experience siting and regulating interstate pipelines for natural gas and oil. I said earlier in my statement that building a new system for CCS transportation could be compared to building the Interstate Highway System. This means creating right-of-ways, landowner rights and liability issues and others. In terms of siting CO_2 pipelines and potentially moving CO_2 across state lines, which priorities should we as policy makers address in order to help CCS projects move forward?

Answer. Your question poses some of the key issues Congress would need to address if it chose to create a comprehensive federal regulatory regime for carbon dioxide pipelines. Among other things, Congress might want to consider what type of ratemaking regime (cost-based or market-based) would be appropriate; whether carbon dioxide pipelines should be common carriers; whether to grant the holders of authorization for carbon dioxide facilities the power of eminent domain; and whether there are specific environmental, economic, or other findings that the siting agency would be required to make in connection with authorizing such facilities.

Responses of James Slutz to Questions from Senator Bingaman

Question 1

Many people have called for ''permanent'' storage in introduced legislation (e.g. all the introduced climate bills), but do not go so far as to define permanence. Can any one of you elaborate on a clear definition of ''permanent CO_2 storage'' as it relates to geologic storage?

Answer. There is currently not a precise definition of ''permanent CO_2 storage.'' Effectiveness of geologic storage is contingent on CO_2 remaining stored underground for a long period of time. A desirable timeframe for geologic storage of CO_2 is on the order of thousands of years or longer. However, effectively sequestering CO_2 for even a few hundred years could provide valuable flexibility in reducing CO_2 emissions and contribute to reducing the costs of mitigating climate change.

Geologic storage of CO_2 has been underway as a natural process in the earth's upper crust for hundreds of millions of years. At geologic time scales, CO_2 forms a natural part of the Carbon Cycle and is necessary for life. In most proposed carbon capture and storage (CCS) storage projects, the goal is to remove CO_2 from the atmospheric portion of the Carbon Cycle, and store it mainly in the subsurface for significant periods. Thus, ''permanent'' must mean no unacceptable change over some long period of time and should provide a reasonable assurance of indefinite storage of the majority of the carbon over a defined number of years (100's to 1,000's). Any leakage from the subsurface to the atmosphere could be mitigated, or the CO_2 could be recovered and stored elsewhere, if deemed necessary. Eventually, an increasing portion of CO_2 stored in the subsurface will be trapped through processes such as formation of minerals, and hydrodynamics with the result that this portion of the CO_2 would be sequestered at a geologic time scale of millions of years. As we gain knowledge from geologic storage projects, a more precise understanding of ''permanent CO_2 storage'' should emerge.

Question 2

What is the appropriate amount of time that CO_2 should be stored in the subsurface, as means of mitigating CO_2 emissions? Will storage times of that magnitude be burdensome on CCS projects?

Answer. DOE and others are conducting pilot scale geologic sequestration projectsto help better understand these questions and others. CO_2 storage in geologic formations should mirror the timescale of oil and gas deposits in formations containing naturally occurring carbon dioxide gas. These formations have held these fluids for millions of years. A desirable timeframe for geologic storage of CO_2 is on the order of thousands of years or longer.

Demonstrating storage over these timeframes should not be overly burdensome.For well-selected, designed, constructed and managed geologic storage sites, the vast majority of CO_2 will gradually be immobilized by various trapping mechanisms and, in that case, could be retained for up to millions of years. Because of these mechanisms, storage could become more secure over longer timeframes (IPCC 2005).

Question 3

Is there any amount of leakage that is acceptable? The IPCC suggests that storage should be on the order of 1000 years in a geologic formation, with lessthan 1% leakage of the volume of CO_2 that is injected over the life of the storage project. Is this a reasonable expectation? How can we enforce such a requirement?

Answer. Yes, this is reasonable to expect. Ideally, of course, no leakage will occur and any leakage that does occur must be taken in the context of how much CO_2 must be sequestered to reach and maintain preferred atmospheric levels. Extensive monitoring and modeling will be implemented at each of DOE's large-scale fieldtests. This monitoring and modeling will occur before, during, and after the CO_2 is injected, for a period of approximately 10 years. This extensive monitoring will result in the opportunity to mitigate any potential leakage.

A natural analog is the million barrels of oil and gas deposits, as well as naturally occurring carbon dioxide gas, that has been trapped in underground geologic formations for millions of years. Furthermore, with proper construction and monitoring,there is a very high probability that the same geologic formations which trap oil and gas deposits and naturally occurring carbon dioxide gas will also help to prevent the significant leakage of carbon dioxide. Sites will need to be chosen carefully, and only the ones with the best geology and proper characterization should be selected for storage. I might add that the United States has a great deal of experience injecting and storing natural gas (in which gas is injected underground during the summerand then recovered to heat homes in the winter) and, though short-term, this experience will also prove useful in developing successful carbon sequestration technologies.

In all cases, best engineering practices will be developed and used when carbon dioxide is injected into geologic formations. The carbon dioxide in the formation is trapped in porous rock, like sandstone. Using a variety of tests and mapping activities should help to identify any fractures in the cap rock and other faults which then can be avoided.

In addition to extensive modeling to determine the potential fate of CO_2, proper site monitoring, measurement and verification (MMV) also can determine if the CO_2 is remaining in the formation. The oil and gas industry has extensive knowledge of monitoring for leaks of various gases from their wells and also for characterizing potential drilling sites for hydrocarbon recovery. This technology along with others being developed through research can be utilized for MMV to ensure that CO_2 leakage is not an issue. Enforcement via continuous monitoring technology may be implemented through a variety of potential regulatory or other legal frameworks that are currently under development.

Question 4

Is the DOE comfortable with the EPA leading the interagency task force described in S. 2323 or would you suggest another organization as the lead agency?

Answer. DOE is comfortable with the EPA leading the interagency task force described in S. 2323. Over the past several years, EPA has been coordinating with DOE. As DOE has developed a Carbon Sequestration Technology Roadmap for the development and deployment of this technology, EPA has been working to design an appropriate management framework for geologic sequestration. By engaging in DOE's R&D program early and working with stakeholders on all sides of this issue, EPA is well-positioned to help in the permitting of future carbon dioxide underground injection wells.

Question 5

As I understand it, one of the chief benefits of the FutureGen approach was that it would demonstrate an integrated design, optimized to maximize CO_2 capture and overall plant efficiency. How are you going to insure we receive this same benefit from a substantially smaller federal investment in a commercial facility?

Answer. The FutureGen program remains a vital component of the Administration's plan to make coal a part of a cleaner, more secure energy future for America. The Administration is restructuring the FutureGen program to accelerate commercial use of carbon capture and storage technology and expand the program from one project to multiple demonstration projects. The 2009 budget request more than doubles funding for FutureGen, from $74 million in 2008 to $156 million in 2009.

Rather than investing in the total cost of an experimental facility integrated with carbon capture and storage, the revised FutureGen approach will invest in the carbon capture and storage portion of commercial power projects, capturing and sequestering at least double the amount annually compared to the FutureGen concept announced in 2003. This will limit taxpayer's financial exposure to only a portion of the incremental cost of the carbon capture and storage portion of the plant. Furthermore, this new approach will allow us to accelerate nearer-term technology deployment in the marketplace faster than the timetable for the previous approach. In order to be successful in competitive power markets (not to mention in the Department's competitive proposal evaluation process), the underlying power plant projects will still need to be efficient, competitive and environmentally sound. The original FutureGen approach was targeting a CO_2 capture and sequestration level of approximately 90 percent, and that level is also specified in the Request for Information for the re-structured FutureGen program.

Question 6

I have heard estimates that including large-scale carbon capture and sequestration on a typical power plant will increase costs by roughly a third. What assurance do you have that the amounts you propose to distribute under this program will be sufficient incentives to lead to commercial-scale demonstration of the technology? Will other federal incentives be available to the applicants, and are more necessary?

Answer. Approximately thirty commercial Integrated Gasification Combined Cycle (IGCC) projects are in various stages of planning, permitting and design across the Nation, which is evidence that a commercially viable basis for IGCC technology already exists. Some are stalled because of uncertainty regarding CO_2. Federal incentives, such as loan guarantees and tax credits, are awarded on a competitive basis and may also be available to some of these projects. Federal funding for FutureGen demonstration projects will help pay for the CCS part of some of these projects this provides additional incentives for such projects. Responses to FutureGen's Request for Information are due by March 3, 2008, and should provide further information on how to structure the FutureGen solicitation to provide sufficient incentives for demonstration projects. Nothing precludes FutureGen applicants from applying for other Federal incentives, such as loan guarantees or tax credits. We have considered the need for further incentives, but believe that none are necessary at this time.

Question 7

In recent months we have seen proposed commercial IGCC plants significantly delayed or cancelled. What assurance do you have that there will be sufficient commercial interest in building these plants to give us the demonstrations we need?

Answer. At the present time over 30 Integrated Gasification Combined Cycle (IGCC) power plants are in various proposal stages and major barriers to their deployment include the uncertainties regarding future CO_2 emissions regulations and the actual costs of constructing and operating IGCC-Carbon Capture and Storage (CCS) power plants. The restructured FutureGen is designed to help understand, address, and solve technical, siting, permitting, regulatory, and fiscal aspects of CCS deployment in various commercial settings. Through its Request for Information, DOE expects to identify the power producers who would consider participating in the revised FutureGen initiative.

Question 8

The 4 phase-3 large-scale CO_2 sequestration tests that have been awarded thus far are all expected to inject less than 1 million tons (approx 500,000) of CO_2 per year—will there be an effort to increase those amounts so that we can have information more in line with that FutureGen would have produced?

Answer. In addition to the four large-scale tests awarded to three of the Regional Carbon Sequestration Partnerships (RCSP) in October 2007, a fifth test was awarded in December 2007 to a fourth RCSP. Three of the tests (in the Alberta Basin, Lower Tuscaloosa Formation, and Entrada Formation) individually are expected to inject at least 1 million tons of CO_2 per year for at least one year. Two other tests (in the Williston Basin and Mount Simon Sandstone Formation) will inject greater than 1 million tons in total, though at a rate of less than 1 million tons of CO_2 per year. The injection rates will be at a scale that demonstrates the ability to inject and sequester several million metric tons for a large number of years. This operation at commercial-scale is as significant as that of higher injections of 1 MM tons per year. It is our intention to confirm the design of these injections, including the applicability of the injection scale proposed for the demonstrations to operations at commercial scale, in a March 2008 technical peer review.

Question 9

The competition for FutureGen between Texas and Illinois led both states to examine the policy framework that would be necessary for CO_2 sequestration. How will the new program create similar incentives for states in which theprojects will be located? What can we do here to accelerate this deployment?

Answer. There are major technical and regulatory hurdles to overcome before coal with CCS can be commercially deployed, however it is in the best interest of states to adopt a posture that would help enable ultra-low emission integrated gasification combined cycle (IGCC) plants with CCS, like FutureGen, to provide stable power supplies at affordable prices.

FutureGen will provide early CCS demonstration experience in a commercial setting, which is aimed at accelerating deployment and advancing carbon capture policy. The revised approach would sequester at least double the amount of CO_2 than the previous approach, generate enough electricity per plant to power 400,000 households, and have the potential of demonstrating CCS in multiple states. FutureGen will help establish commercial feasibility and formulate a model that industry could use to deploy commercial-scale plants that each sequester at least one million metric tons of carbon dioxide annually.

Responses of James Slutz to Questions From Senator Dorgan

Question 1

It seems to me that we need to much more quickly begin establishing and defining the ''rules of the road'' when it comes to carbon management. As we begin to unlock the opportunities for capturing, moving and storing larger amounts of CO_2, it is fair to say that the federal government will likely play a greater role. It will be better if we begin to better define appropriate roles for local, state and federal government. What are the most critical near-term issues that your agency can address so that developers can begin demonstrating CCS projects?

Answer. The most critical near-term issues DOE can and is addressing through its research, development, and demonstration carbon capture and storage (CCS) program are the development of technology for CCS, which in turn will advance public acceptance of CCS as a technology for mitigating greenhouse gas emissions. Testing the storage of CO_2 in deep saline formations and depleted oil fields will enable representatives from industry, states, the U.S. Environmental Protection Agency, and others to learn from the field demonstration of site characterization, operations, and closure which will lead to the development of best management practices for all aspects of CCS projects. These field activities, which DOE is supporting on a cost-shared basis, are critical in the deployment of these technologies and will provide key information in the development of regulations and policies to support CCS. The near-term successes of the field activities will help to support demonstrations planned under the restructured FutureGen Program [at 1M tons CO_2/yr, FutureGen is about the same scale as the field tests].

Question 2

Creating an infrastructure to capture, transport, store, and monitor CO_2 will take greater federal resources including staff, technology and other elements. Do you think your agency is well-equipped to begin undertake this enormous challenge?

Answer. The DOE has a dedicated interdisciplinary team, working to develop and demonstrate technologies for carbon capture and storage (CCS) and provide support to other agencies in the development of regulatory structures for injection. Additionally, DOE works with other countries through the Regional Carbon Sequestration Partnerships (with over 350 entities involved) and the Carbon Sequestration Leadership Forum. DOE also has the ability to add resources from the nineteen national laboratories and enter into cooperative agreements with industry and other research institutions when necessary.

Question 3

DOE has been implementing the Carbon Sequestration Regional Partnership program to study carbon management in different regions of the country. Also, your agency has been working to identify storage sites for CO_2. I think this work is important and should continue to help us evaluate where the opportunities for CCS demonstrations could take place.

I do believe that we can find cleaner and more efficient ways to utilize our coal resources. However, we must do so in a way that does not jeopardize are ability to generate base load power. Your agency has several different areas working on different elements of carbon capture and storage.

Could you provide me with an explanation of how the Department of Energy is coordinating the different research, development, demonstration and deployment program areas working on CCS in order to deploy the technology more rapidly?

Answer. There are several elements of DOE's Sequestration Program for carbon capture and storage (CCS). They include the R&D which fund basic and applied basic research for carbon storage and capture technologies. The projects funded through these R&D programs represent innovative approaches that can significantly reduce the cost and demonstrate the safety and effectiveness of CCS. The second part of the program consists of large-scale CO_2 injection projects, which are designedto take the technologies developed in the R&D programs and deploy them in the field through programs like the Regional Carbon Sequestration Partnerships. This part of the program is also responsible for developing the infrastructure technologies and information, such as CCS best practices that could help form a basis for regulations, for CCS deployment through the involvement of representatives from industry, non-governmental organizations (NGOs), universities, and Federal and statepartners. The final piece of the DOE CCS program will be implemented through the clean coal demonstration program (such as the Clean Coal Power Initiative and FutureGen demos), which will take the technologies developed from the R&D and large-scale injection projects and implement these in full-scale power plant demonstrations that include CCS. Early commercial deployment of plants with CCS can benefit from FutureGen demos and other deployment incentives. The SequestrationProgram, which is managed by the Office of Fossil Energy, also coordinates with DOE's Office of Science to enhance the scientific learning and understanding in the field demonstration projects.

All of the projects awarded through these DOE programs are based on cooperative agreements with industry and/or research institutions. Therefore, the success of these programs depends upon the success of our partners and DOE's continued efforts to promote technology transfer.

DOE is also supporting working groups through other Federal agencies, non-governmental organizations (NGO), and industry that are working to develop regulations and liability frameworks, and to educate stakeholders about the benefits of CCS.

Responses of James Slutz to Questions From Senator Landrieu

Question 1

I learned that the Department of Energy has decided to alter course on its ''FutureGen'' project. In the Request for Information that you released yesterday, you describe that your new approach will target IGCC plants to demonstrate carbon capture and storage technology. In Louisiana, we have saline storage formations, and we have a 300 megawatt power plant coming online. At present, it will be fueled by Petroleum Coke, but it is not currently slated to be an IGCC plant. Will the Department keep an open mind about selecting plants that may not be IGCC equipped, but that are nonetheless capable of capturing and storing their carbon?

Answer. Yes, alternatives to Integrated Gasification Combined Cycle (IGCC) will be considered. The Request for Information does seek comments on whether the revised FutureGen approach should allow for advanced coal-based technology systems, other than IGCC, that would meet the stated performance requirements for FutureGen (e.g., approximately 90 percent CO_2 capture and storage, 0.04 lbs/millionBtu SO2 emissions, less than 0.05 lb/million Btu NO_X emissions, less than 0.005 lb/million Btu particulate matter emissions, and greater than 90 percent mercury removal).

Question 2

Additionally, your Request for Information states that the Department will contribute the incremental cost associated with adding CCS technology to the facilities power train. Would the DOE cover the costs associated with compressingand transporting the CO_2? Are these grants only intended to cover capital costs, or will they cover certain qualified operating costs as well?

Answer. DOE will contribute a portion of the incremental cost associated with adding CCS technology. Based in part on input obtained through the Request for Information, DOE will determine which incremental costs are eligible for cost-sharing, such as compressing and transporting the CO_2 and certain operating costs. Thedetermination of which costs will be eligible for cost sharing, particularly for any equipment that might be shared between the power plant and CCS technology, will be articulated in the formal solicitation.

Responses of James Slutz to Questions From Senator Cantwell

Question 1

The Energy Bill (Title VII, Section 702) passed in December directs the Department of Energy to conduct not less than 7 initial large-scale sequestration tests to study and validate commercial deployment of technologies for CO_2 capture and sequestration. Seven regional partnerships have been identified and arecurrently entering Phase III of their projects. I understand the Department has currently awarded 4 of the 7 partnerships with Phase III funding. What is your path forward to provide funding for the final three in 2008?

Answer. DOE has made awards for five large-scale tests to four of the Regional Carbon Sequestration Partnerships (RCSP) for Phase III Large Volume Sequestration Testing. Depending on the results of a scientific needs assessment being conducted in FY 2008 and the ability of additional project proposals to meet thoseneeds, additional projects may be awarded in FY 2008 or FY 2009. The remaining three Phase III projects are in the process of being evaluated. The evaluation process requires finalizing the technical scope of the project along with undertaking a scientific evaluation and cost analysis of the proposed projects to verify their appropriateness within the overall objectives of the Sequestration Program. Independent cost verification is being undertaken by DOE to ensure the project costs are adequate prior to award. Independent cost reviews of the projects that have received awards have been completed. An independent technical review will be conducted at the end of March 2008. This technical review, conducted by an internationally renowned group of experts, will compare the proposed test plans against the program needs and that required for proper scientific evaluation in order to develop an integrated portfolio of robust tests. DOE is conducting reviews and plans to evaluateaward of the remaining RCSP Phase III Projects based on the results on the scientific evaluation. The estimated time-frame for evaluating the remaining awards is the summer of FY 2008. The Sequestration Program budget is available to fund these awards.

Question 2

Which partnership takes into consideration the geologic formations in the Pacific Northwest and what is the status of this project?

Answer. The Pacific Northwest is shared by two Regional Partnerships, the Big Sky Regional Partnership (Big Sky) and the West Coast Regional Carbon Sequestration Partnership (WESTCARB). The Big Sky Regional Partnership is currently evaluating basalt formations in the region. The West Coast Regional Carbon Sequestration Partnership in addition to the West Coast is responsible for working with the states in the Pacific Northwest to characterize the geology and terrestrial sinks inthat region. The WESTCARB project has completed the initial characterization of saline reservoirs and coal seams which could be possible storage formations. The results of this characterization is available through the National Carbon Sequestration Database and Geographic Information System online Atlas at the following site: http://www.natcarb.org These Regional Partnerships are currently in Phase II, undertaking Field Validation Testing in the region, and are two of the awards thatare in the process of evaluation for potential award under Phase III.

Response of James Slutz to Question From Senator Menendez

Question 1

In April, in testimony before this committee, Thomas Shope, Acting Assistant Secretary for Fossil Energy at the Department of Energy, estimated that Carbon Capture and Sequestration technologies would become deployable and available in 2020 to 2025, but that wide deployment for most projects would not happen until 2045. Do you agree with this estimate? If so, won't this be too late if we are going to reduce our greenhouse gas emissions by 80% below 1990 levels by 2050?

Answer. Widespread deployment of Carbon Capture and Storage (CCS) depends on a variety of factors, including success of R&D to drive down the cost of safe CCS, particularly the cost of separating CO_2 from other gases and compressing it (to a supercritical fluid) for injection into geologic formations, and success of demonstration of CCS technologies so that the lowest cost technologies can be identified and commercialized in a timely manner.

The Administration believes that significant reductions in CO_2 can be made through investment in technology that will lead to a fundamental change in the way we produce electricity and power our vehicles. The President's 2009 budget request for research, development and demonstration of advanced clean coal technology, when combined with required private-sector contribution, will approach a total investment of nearly $1 billion. With continued support, it could be possible to significantly advance the timing for full deployment of CCS technologies.

[Responses to the following questions were not received at the time the hearing went to press:]

Questions for Stephen Allred From Senator Bingaman

Question 1

Many people have called for ''permanent'' storage in introduced legislation (e.g. all the introduced climate bills), but do not go so far as to define permanence. Can any one of you elaborate on a clear definition of ''permanent CO_2 storage'' as it relates to geologic storage?

Question 2

What is the appropriate amount of time that CO_2 should be stored in the subsurface, as means of mitigating CO_2 emissions? Will storage times of that magnitude be burdensome on CCS projects?

Question 3

Is there any amount of leakage that is acceptable? The IPCC suggeststhat storage should be on the order of 1000 years in a geologic formation, with less than 1% leakage of the volume of CO_2 that is injected over the life of the storage project. Is this a reasonable expectation? How can we enforce such a requirement?

Question 4

In your testimony you mention that the Dept of Interior will have a critical role in determining how CO_2 is managed on public lands. One area your testimony did not discuss was the site selection criteria that will be necessary in choosing geologic formations suitable for storing CO_2. The USGS employs some of the world's leading geologic experts. In your opinion, would the USGS be a good organization to recommend a set of ''best practices'' for geologic site selection for CO_2 storage? By this I mean that they would not be regulators of the site selection, but instead recommend the technical requirements for safe, long-term geologic storage of CO_2.

Questions for Stephen Allred From Senator Dorgan

Question 1

It seems to me that we need to much more quickly begin establishingand defining the ''rules of the road'' when it comes to carbon management. As we begin to unlock the opportunities for capturing, moving and storing larger amounts of CO_2, it is fair to say that the federal government will likely play a greater role. It will be better if we begin to better define appropriate roles for local, state and federal government.What are the most critical near-term issues that your agency can address so that developers can begin demonstrating CCS projects?

Question 2

Creating an infrastructure to capture, transport, store, and monitor CO_2 will take greater federal resources including staff, technology and other elements. Do you think your agency is well-equipped to begin undertake this enormous challenge?

Questions for Stephen Allred From Senator Smith

Question 1

I want to thank you for your response to my December 2007, letter concerning the pending Memorandum of Understanding between the Minerals Management Service (MMS) and the Federal Energy Regulatory Commission (FERC) regarding wave and current energy projects on the Federal Outer Continental Shelf (OCS). It is my understanding that this issue also came up at the Energy Committee on Thursday. I appreciate the actions taken by MMS, but I remain concerned that the potential for environmentally-friendly wave energy development will continue to be delayed on the OCS. Can you tell me when these proposed regulations, which have yet to be published for public comment, will be finalized? In the current draft of the agency rulemaking, do you address the issue of jurisdiction between MMS and FERC, which I understand would have been addressed in the MOU? If not, why can't MMS sign the MOU in order to provide regulatory certainty on agency jurisdiction now, and seek to amend the MOU if the final regulations require such a modification?

Question 2

You stated in your letter and at the Committee hearing that you had been asked by the Committee not to sign the MOU. It is my understanding that such a request was orignally made when there was Senate-passed language in the 2007 energy bill that would have specified that FERC did not have jurisdiction over kinetic hydropower facilities located in the OCS. However, that language was not included in the final version of the bill, which is now P.L. 110–140. Can you tell me why, in the absence of this language, it wouldn't be helpful to those seeking to develop projects on the OCS to provide immediate clarity concerning the regulatory roles and responsibilities of the two agencies?

Question for Stephen Allred From Senator Wyden

In my State of Oregon and in the State of Washington, western coastal basins offer potential carbon sequestration opportunities. Promising basins include the Puget Trough, Tofina-Fuca Basin, West Olympic Basin, Whatcom Basin, and Willapa Hills Basin in Washington, and the Astoria-Nehalem Basin and Tyee-Umpqua Basin in Oregon.

Furthermore, Oregon is one of the states that is part of the Big Sky Carbon Sequestration Partnership (BSCSP). Their vision is to prepare its member organizations for a possible carbon-constrained economy and enable the region (Montana, Idaho, South Dakota, Wyoming, eastern Oregon and Washington, and adjacent areas in British Columbia and Alberta) to cleanly utilize its abundant fossil energy resources and sequestration sinks to support future energy demand and economic growth. The BSCSP will achieve this vision by demonstrating and validating the region's most promising sequestration technologies and creating the supporting infrastructure required to deploy commercial scale carbon sequestration projects. BSCSP has the goal of developing an infrastructure to support and enable future carbon sequestration field tests and deployment throughout the BSCSP region.

This technology is extremely attractive in assisting to address climate change issues; assuring the environmental acceptability and safety of carbon dioxide (CO_2) storage in geologic formations and determining that CO_2 will not escape from geologic formations or contaminate drinking water supplies are major concerns. Much research is needed to better understand and characterize sequestration of CO_2 in geologic formations; I understand that researchers are building on the significant baseline of information and experience that exists.

Question 1

During the second panel discussion of the hearing, Mr. Allred of the Department of the Interior discussed issues pertaining to land leasing requirements and mineral rights, as well as carbon dioxide ownership and eminent domain. While this discussion was informative, no resolution was provided on the subject matter. Mr. Allred, can you provide details on how carbon capture, transport, and sequestration, as well as the creation of carbon dioxide transport pipelines will impact private land owners' property and mineral rights in Oregon?

And what are the federal land right impacts of carbon capture, transport, and sequestration on public property such as national forests and those supervised by the Bureau of Land Management?

APPENDIX II : ADDITIONAL MATERIAL SUBMITTED FOR THE RECORD

AMERICAN PUBLIC POWER ASSOCIATION,
Washington, DC, January 28, 2008.

Hon. JOHN KERRY,
U.S. Senate, 304 Russell Senate Office Building, Washington, DC.

Dear Senator Kerry: I am writing to express the American Public Power Association's (APPA) support for Section V of your bill (S. 2323) that establishes an interagency task force to develop regulations providing guidelines and practices for the capture and storage of carbon dioxide.

As Congress continues to debate climate change, one of the most frequently discussed technologies is that of carbon capture and storage. While this may be a viable option to address climate change, there are major challenges that must be overcome, both technically and in public policy, before widespread commercial-scale carbon capture and storage can be achieved. APPA believes your bill is a step in the right direction to overcoming these challenges.

Again, thank you for your dedication and hard work on this matter and we lookforward to working with you as your legislation moves forward. Sincerely,

MARK CRISSON,
President & CEO.

Statement of Robert J. Finley, Director, Energy and Earth Resourcescenter, Champaign, IL, on S. 2323

The Illinois State Geological Survey is one of the largest and most diverse state geological surveys in the United States. We have been researching carbon sequestration in the Illinois Basin of Illinois, southwestern Indiana, and western Kentucky since 2001. In 2003, we submitted a successful competitive proposal that began our work as a lead agency for a Department of Energy (DOE), Regional Carbon Sequestration Partnership. Phase I of that work was completed in 2005, and a Phase II program was awarded in 2005 that runs through 2009. Our

Phase III large-scale sequestration test was awarded this past December and we will inject one million metric tonnes of carbon dioxide (CO2) into a saline reservoir with the collaboration of the Archer Daniels Midland Company who is providing the site and the high-purity carbon dioxide for the test. Other tests we are conducting include injection into mature oil reservoirs and into coal seams to evaluate enhanced oil recovery and the ability of coal to adsorb CO_2. With these efforts we have established a program to address the major potential reservoirs for carbon sequestration, as have the six other Regional Carbon Sequestration Partnerships in their respective regions around the country. The existence of these efforts leads me directly to offer comments on S.2323.

Section 3 of this bill is essentially duplicative of the work that the DOE Regional Carbon Sequestration Partnerships are conducting as part of their Phase III efforts. The projects already planned and underway involve 1 million tonnes from a variety of commercial sources. For purposes of sequestration, and assessing the safety and effectiveness of the process of isolating CO_2 from the atmosphere, the Partnership tests are addressing the very issues this bill proposes to address. Given the urgency in developing responses to climate change, I cannot support initiating a new effort that duplicates work that is already in place with similar goals, volumes of CO_2, and geographic diversity.

I would also call your attention to Section 6. A significant amount of research has already been conducted or is now underway with respect to carbon capture by the Office of Fossil Energy within DOE. S.2323 seemingly does not provide for coordination between Office of Science and the Office of Fossil Energy with respect to the work that already has been accomplished. Thus, the potential for duplicative work again arises.

It is important that Section 7 of S.2323 recognizes the work on capacity assessment that has been completed and is now being updated by the DOE Office of Fossil Energy. The initial methodology is already undergoing modification and refinement. However, some of the work proposed in this section is likely to be duplicative and the required coordination provisions should be strengthened. Certainly, some detailed, recent assessments of the volumes of oil incrementally recoverable through CO_2 enhanced oil recovery have been made on a basin scale, published, and presented around the nation. With regard to the Geological Verification provision of Section 7, I would suggest that this element be dropped. The authorized funding will not cover any useful new drilling in its entirety while providing for the other aspects of this Section. It would be far more effective to focus on partnerships with existing drilling efforts in order to specifically cofund data collection (coring, advanced well logging, and similar) but not make any contributions to actual drilling expenditures or footage rates.

Senator Bingaman and Members, I appreciate the opportunity to submit these comments to the committee and would welcome any follow-up communications that may be useful to you.

End Notes

* Figures 1–2 have been retained in committee files.
** The graph has been retained in committee file.
*** Document has been retained in committee files.
[1] Total costs of CCS varies substantially by source of CO_2—to upwards of $70/ton—and even across proposed gasification projects because of variances in each process. This figure represents an estimate of the lowest-cost industrial-sourced CO_2.

INDEX

A

B

C

D

E

F

G

H

I

J

K

L

M

N

O

P

Q

R

S

T

U

V

W

Y

Z